MW01031287

VAGABONDS

VAGABONDS

Tourists in the Heart of Darkness

NICK BROKHAUSEN
AND JEFF MILLER

Published in the United States of America and Great Britain in 2021 by
CASEMATE PUBLISHERS
1950 Lawrence Road, Havertown, PA 19083, USA
and
The Old Music Hall, 106–108 Cowley Road, Oxford OX4 1JE, UK

Hardback Edition: ISBN 978-1-61200-995-7
Digital Edition: ISBN 978-1-61200-996-4

A CIP record for this book is available from the British Library

Printed and bound in the United States of America by Sheridan

Typeset by Lapiz Digital Services.

For a complete list of Casemate titles, please contact:

CASEMATE PUBLISHERS (US)
Telephone (610) 853-9131
Fax (610) 853-9146
Email: casemate@casematepublishers.com
www.casematepublishers.com

CASEMATE PUBLISHERS (UK)
Telephone (01865) 241249
Email: casemate-uk@casematepublishers.co.uk
www.casematepublishers.co.uk

Front cover image: Shutterstock/Vlad Vahnovan

The wind whispers in my ears, and the moonlight lights my way,
My past is behind me, abed and tucked away with my fears.
The rails will carry me, to tomorrow and whatever might lay.
Vagabond, Vagabond, follow your heart til the end of your years.

—Nick Brokhausen

Contents

These are the somewhat twisted memoirs of two former military nomads and their friends, as well as a few of their enemies. Their lives are entwined with history, having been witnesses and participants. It is not a travelogue, nor a confession, and should not be viewed as a guidebook for the soul.

Every chapter/episode actually happened, maybe not exactly as described herein but close. There have been changes made for story purposes, some to make the narrative flow more smoothly, some to dodge possible classification issues. Some characters have been changed and sometimes two or more people have been consolidated into a single character. Names have been changed, not always to protect the innocent, but the underlying story is, for the most part, what really happened.

The story is told in Nick's voice but it is a compilation of both of our stories. Often we were together, occasionally we were on our own, but between us we managed to cover a pretty large swath of this planet. We didn't write this book to make ourselves out as heroes, quite the contrary. We wanted to memorialize what a life like ours was really like. A lot of confusion, a lot of humor, a lot of broken dreams, broken promises, broken marriages, and the occasional triumph. It isn't like Hollywood, that's for damn sure. It is much more complex than Hollywood could ever imagine.

CHAPTER ONE

Birth of a Notion

In our journey through this life, we are faced with choices and situations, which array themselves against the backdrop of consequence and the bizarre. Our fate plays itself out while we, impressed with our ability to accomplish our dreams, stumble forward like some Pavlovian experiment, stoically taking our wounds and suffering as fare for the trip.

Once one has survived the ultimate contest between skill, luck and happenstance which defines combat, everything else seems easy and within reach. In my military career I had participated in some of the most dangerous and daring operations the United States had undertaken as a nation. During that time I met some of the most incredible and intelligent people, who encouraged me in my madness and tutored me in the skills to survive and prosper in my endeavors. This is the beginning of my life journey, and that of my closest comrades, after we left the womb of the military. It shall be their story, for they deserve the recognition, good or bad …

I am swimming up from the depths of a catatonic sleep, still in that nether land of some feverish dream. I crack one eye open and try and focus on my surroundings. I am on a journey of awakening, but the way my head feels I may have premature expectations. I manage to get both eyes open and adjusted to the dim light. I hear a chirping sound off to one side, and as I move my head to better focus on the source, the first thing that looms into view is a large male raccoon who is busy chewing on what appears to be a half-eaten slice of pizza. I still think I am in the clutches of the dream state. I lay there like a curare victim, paralyzed and unable to do anything more than drool.

My God this thing is huge. It must weigh 40 pounds and has that sleek look of the urban garbage-can aficionado. It keeps staring at me as it chews the pizza and makes little chirping noises as if we were breakfast chums. I am feeling around as I try to get some distance from it, and my hand grasps what is most assuredly a pistol. It's tangled up in the covers and sheets which are trapping me as well. I'm effectively trussed up like a Thanksgiving turkey ready for the oven. I keep trying to free the pistol and get it up in case this apparition from the Bullwinkle show shows any sign of being hostile.

I am surveying my surroundings at the same time since I can't remember where I am. How did I get here? And which of my friends were involved? At least they left me a gun. I get the gun free. Ah, it's a Browning Hi-Power, and it's mine. I start looking for my clothes as I manage to get to the head of the bed without being swarmed by my breakfast date. I am slowly getting up as I spy my discarded clothing. I move over and start pulling together the appropriate rumpled ensemble, all the while making soothing noises to the fur ball with his pizza-box cuisine. He has finished the contents and now he is beginning to chew the cardboard—lovely.

I am beginning to recognize my *where*. It's the spare bedroom of Jay's humble abode in Ayer, with the décor of a military surplus warehouse crossed with a speed shop. I also remember the raccoon's name: it's Rocky, and Rocky has now decided that I might know additional food sources and follows me out of the bedroom and down the hall to the kitchen.

I remember now how I got here. I had just processed out of the army, with my illusions shattered and a bad taste in my soul for the future. My last gasp at trying to reconcile 17-plus years in Special Forces had been the Iranian mess. Charlie had spun up the newly minted Delta Force, which was brimming with former Son Tay raiders, and had trundled it forth to stab the Ayatollah minions in the heart. A lightning-fast strike force with overwhelming fire support, with thousands of moving parts, had been cobbled together. It was a good plan and only had one fatal flaw. The umbilical cord stretched all the way back to the White House, and the go/no go button was in the hands of the Georgia Mafia surrounding the president. The ground commander had no latitude to adjust, adapt, or do anything other than answer the satellite phone. Throw together an amazing chain of events that unfolded on Desert One and the whole thing disintegrated. The subsequent second attempt and the political roller coaster surrounding it had convinced me that there was no adult supervision higher than the group level. I was thoroughly disgusted with the army and the government in general.

My disillusion was to the point that I needed to reinvent myself. Thus, here I was out of the service and looking at my prospects. Rhodesia and the African market for my skill sets were winding down and the Banana Wars had just begun to spin up. I was tooling an idea over with two other Special Forces types who were getting out. We were all in Massachusetts, where we had served together in the 10th Special Forces Group.

Jeff Miller is a gifted, if not totally twisted, fellow traveler. I suspect that he, like Doberman Pinschers, has trouble with his brain not fitting in his slower-growing skull. Jeff is short, compact, and a direct descendant of the Miller clan who rode with Frank and Jesse. In fact, his great-granduncle Clel was gunned down on the streets of Northfield. I had found an old picture taken after his esteemed and newly deceased uncle had been put on display in front of the general store. Draw in a pair of military glasses, and zip, perfect resemblance. Jeff wears correctional glasses, and they make him look and see better. Without them he resembles a cross-eyed wolverine.

Penguini, the third part of the troika, and I had been in Berlin together. He was retiring out as a master sergeant, whereas Miller and I had pulled the cord short of 20 years. Penguini is an extremely talented martial artist. In fact, his brother was one of the Memphis Mafia around Elvis Presley. He and Red and Sonny West were the inner core around Elvis. His brother, who we and others refer to as the Grand Master or just the GM, is tall, imposing and deadly, whereas Penguini is compact, jovial, and deceptively harmless looking. His nickname of Penguini is based upon a notorious picture that found its way onto the wall of many a *Gasthaus* and bar in Berlin. We had shared adventures, scrapes, and advanced fibbing, as well as a couple of women during our time in Berlin. I knew him as well as I knew my dark side.

The three of us have decided to put together a training company which we will use to teach SWAT teams and other interested parties the fine art of counterterrorism while attempting to possibly insinuate ourselves into some of the darker corners of our government. Mind you this is before the tsunami of private security and military contract groups that would emerge 20 years later. At that time private security was dominated by Wackenhut and Kroll Associates.

The elite of the elite in the security industry were former FBI, with a mixture of flotsam and jetsam from the alphabet-soup crowd dominating the upper echelons. We were aiming at a narrow slice of that pie. Little did we know that the lack of even basic business skills would flame out our ambitions later. As it was, we were ready to take on the world of high-speed action, using our skills and knowledge.

We had our first customer in the form of the International Association of Chiefs of Police, to deliver SWAT seminars to selected customers drawn from all over the US and the world. These would be three-day seminars in locations around the US. Chief Quinn from Newton had been a sponsor of our absorption into respectability and eventually led to our association with Frank Bellotti, the attorney general of Massachusetts. Frank was a UDT veteran from the Korean War and kept a beer tap in his desk. He was also the most powerful AG in the country. This man was assisting in finding us work and keeping a watchful eye that we didn't surrender to the dark side.

Penguini and I had been using alternative housing since our budget was in the red. We had commandeered a demo well in a bridge abutment, which we tricked out with castoffs like a couch, rug, bunks, and a hot plate, then pulled power from the building next door with a long patched-together extension line. It was a state building so we surmised the load loss would go unnoticed. We came home one day to find the AG and his imposing chief of detectives sitting on our couch, drinking our beer and noting the number of felonies attached to our choice of abode.

Our next domicile after Frank found out about our fortress of solitude was the chief of detectives' empty house in Dedham. Great place but no heat in February, which made hygiene an exercise in standing in a tin tub, then pouring hot water from a pot, heated on the kitchen range; then soaping quickly, whilst having the other guy hold a gun on you to encourage you to pour a pot of cold for the rinse. After a couple of months, we finally secured a flat across the commons from the state house and actually were anticipating living like normal people. We had come up to Ayer for a real hot shower and alcohol before we could move in.

I am starting to recover some of the events from the night before, and they're not pretty. We had been club crawling in Leominster and Ayer. We had gathered old acquaintances as the night wore on. By midnight our prospects for attracting the opposite sex had dwindled to the desperate, so we had finished up the night at Jay's hill-side mansion for the heathens. I remember this as the raccoon waddles into the kitchen. I am busy trying to coax some coffee out of the Mr. Coffee by adding water to the tarlike remnants then heating it. Rocky squats on his hind legs and pushes his hands forward in supplication, all the while chirping some coon dialogue at me.

I hate this raccoon and its owner. They are a visual reminder that the original idea of seeking change had been to progress, not regress. I am still not quite sure what the real world will offer—so far if the last few weeks had been a harbinger, it has been like a military field trip on acid. We are the eternal optimists. No matter the obstacle, we can adapt and overcome. Except for Penguini that is, who is convinced he is dying of something and is running around with eye color charts to read liver health and getting high colonic enemas to de-toxify. Knowing his past, he will need to change the mixture to quicklime and sulfuric acid.

I hear stirrings from some of the scattered bodies lying around on couches, corners, and one in the bathtub. All fine examples of "rough men standing ready to visit violence on our enemies." Rough as a descriptor being the prime word. It's Sunday and this place reeks; meanwhile, the raccoon is licking someone's face, waking him to a shrieking reality. I'm thinking that Sunday is off to a fine start: with religious services being optional, most of the survivors are looking for a cold beer. The refrigerator has already been looted and someone else remembers a stash in a cooler outside. The coffee is pure caffeine and tastes like asphalt, but I am awake. The phone rings and I pick it up. It's Miller and he is on his way over here to link up for breakfast. He has the details of our first contract. I am great with that idea. With this masked night denizen finishing off any scraps left over from last night and the fridge robbed like the 3:10 to Yuma, eating out is haute cuisine. I make my way to the door, running into Jay on the way. That boy has so much body hair he looks like a bigger version of his little pal Rocky. The raccoon has better breath though. He burps a query of where I'm going, so I tell him someone puked in the john and I'm going to use a tree.

As I clear the house, I see him peeking at me out the veranda. He suspects that I am going after food, and since I know his spending habits, if it is after the 10th of the month, he is broke, he's looking for someone generous and it's not me. I walk out to the driveway as Miller pulls into the entrance. He has the Penguini with him. I slide into the car and we make off for Denny's. Perfect for this morning, a good wholesome breakfast with a gallon of coffee. Loud enough that people don't eavesdrop easily and not yet full of the party survivors from last night.

Miller is excited which means that there is a hint of money on the air. The Penguini is equally excited. He has a nose for money, especially if he gets some of it. Jeff is his usual laid-back self but pensively atwitter with the plan. He has been romancing the IACP through his contact with a large security firm. He has been "conventioning" with that crowd and has somehow talked them into giving us the contract to present all their SWAT seminars. The students will come from departments all over the US and internationally. I'm starving so I am only getting half of his debrief as he wings his way through traffic like he has a full cruiser with the lights on. Guaranteed number of classes, multiple cities, travel expenses in the budget and we get to flog a couple of our own products. Sounds great, much too perfect, I am sure there is a hook in there somewhere. The money is barely enough to keep heart and soul together, but it will lead to other things. So, we bite and bite hard. Our real interest is selling our design for an armored shield that can be used for a variety of situations where you don't want to be the bleeder.

Jeff is the perfect obtuse for our troika. He has the smarts, the photogenic face for the front and he is a snappy dresser. Sort of a Hannibal Smith cloned with Face and Murdock. A hybrid if your taste runs to waters dark and deep. He also has the memory of a main frame, and visual accents like cuff links. Additionally, he is far calmer than the Penguini and myself. I know from past experience that Penguini will always have a preplanned escape route, so I stick close to him and act as interpreter between him and the Brain. Taken together we might pass as a whole human.

We are to present our first set of the program next week in Newton. Chief Quinn has cobbled together principal players in this newfangled SWAT concept. We know a good number of them and have worked with them in the past. Most are the cookie-cutter copy of beat cops and are most decidedly Irish. These are third- and fourth-generation cop families, churning out generations of the thin Blue Line. It's almost poetic; until you drink with them. We are acceptable since we are a similar gene pool.

We arrive at the Denny's and soon are happily awaiting the feast, going over the minutiae and structuring the seminar, breaking down the instructional content. I hear more people coming in behind us and ignore it at first. Then I hear the distinctive chirping of that forest rat behind me. I turn and there is Rocky squatting on the floor with his two little hands outstretched again in supplication. He is on a dog leash and at the other end is his owner, the ogre. Jay doesn't miss a beat: he slides in next to me and asks if we have already ordered. I know this drill. He will order when the waitress comes back, and she will assume we are all on the same bill. Rocky has climbed up in his lap and is having a grab fest at anything loose on the table.

The waitress arrives and immediately tells Jay that dogs, cats, hobos, and critters are not allowed. Jay demands to see the manager. The manager comes out and tells him to take his critter outside. There is a spirited conversation about rights, animal rights, health codes etc., but the manager is unmoved by the schmooze. Jay gets up, exits the café with his fur-covered pal, and walks over to his Cadillac. We can see the whole event transpire. He can't let Rocky in the car because the little beasty will have chewed the interior into scrap by the time breakfast is over. So he grabs him and walks over to the trunk and jams him in with the spare tire and car parts. He is soon back with us and inevitably gets wind of our little enterprise. Jeff has already filled out the instructor list and we placate Jay with promises of including him in some future event. Jeff has though added one person who is not prior military. This is in the form of Max. Max was a castoff from the agency. Yes, that one. He had been an advisor to the mujahedeen in their fight against the Soviet Union. He had been terribly efficient and had taken to the locals and culture with a vengeance. Unfortunately for the Langley crowd, he had gone native. When they sent the stand down order he went back into the hills with his adopted brethren. They had eventually managed to pull him out of the Hindu Kush and back to Washington for reassignment. He did some work in Washington for a while which consisted mainly of taking Arab officers to the various strip joints and occasionally surreptitiously filming their liaisons for future use as a lever.

He got his nickname because he looks exactly like Max Headroom on MTV. Max is looking for extra income because his latest girlfriend chucked him out on the streets. We are going to meet him and go over the tasks. He hasn't had formal instruction in class presentation or content, but we assume that if the agency hired you it stands to reason that you have some skill sets. Our assumptions will prove to be fragile at best.

We get up to go and of course Jay has to go and check on Rocky before the check arrives. We are dividing up the bill when I see him flip open the trunk lid. He stands there with the few food scraps that he had filched from the table, looking down and talking to his little pal. We can't see the raccoon but evidently confinement doesn't agree with him as he launches himself and lands on Jay's chest and claws his way up to his shoulder administering some bite complaints along the way. Jay goes berserk and flails around the back of the caddy with his self-animated raccoon stole. He finally wrestles him into the car and tight reins him, then squeals off.

Miller, aka the Beast, has fallen into the role of *The A-Team*'s Hannibal Smith and within the next week we have assembled our group and are ready to do our first seminar. We have all rehearsed our classes in front of the others. All, that is, except

Max. He was in Washington, burdened with some important task that his masters tapped him for at the last minute. After being fully versed in his past history, I am sure that this is not the truth, for whomever his managers are, they avoid him like the career-wrecking adrenalin junky that he is.

The day of the seminar we are primed and ready to go and Max is a no-show. We are awaiting his arrival and plotting what to do with the body if and when he does show, when we get a call from the airport in Boston. His plane was late, he forgot his homework, he ran out of gas, all blarney. He more than likely spent the night in his car and overslept. Chief Quinn ends up sending a squad of cars to pick him up with sirens and all. He arrives and breezes past us like the rabbit in Alice in Wonderland and begins his presentation.

He has crib sheets that he is reading from in the most monotonous, boring drone we've ever heard. Everyone's heads are lolling, and some are asleep in the first 10 minutes. He never looks up and continues his droning monologue. He has eventually run out of crib notes, which were out of order anyway, and has resorted to notes he has written on the palm of one of his ham-like mitts. There seems to be no end to this. Penguini in desperation saves the moment by setting fire to the trash can in the rear and allows us to get Max off the podium. He was right in the middle of outlining on whiteboard the proper step-by-step execution of a search pattern. Each step represented by some fatiguing dot matrix. We rush him out to the hallway like a bad vaudeville act.

Evidently the farm doesn't process their trainees in any sort of instructional format. Not surprising since their job is to get someone to betray their country, whilst Special Forces is designed to train, equip, and lead guerilla forces and develop an underground. This is why the post office in Langley has a paramilitary division, chock full of former SF. Max unfortunately was from the deceit and betrayal division. But he is full of interesting information. All we need to do in the future is siphon it out of his brain, just don't let him present the class. So Max gets a pass on the first one and we find his niche in our vision. Basically, he is a cataract in the movement, blurred sight.

* * *

We have begun to design and produce equipment to augment SWAT-type operations. The first and best idea is a design for a ballistic shield with a see-through window and a gun port that resembles a Roman legionary's *scuta* or shield. We have the basic design and a prototype finished when the Penguini's older brother the Grand Master calls and convinces us that there are a ton of opportunities in California for developing our training and equipment market. So we pack up and drive to California in our leased Camaro. California is a culture shock but, in some ways, pleasantly so. We go from surviving to surviving with a purpose, and in nicer weather. We establish a new company and combine it with the GM's karate empire, and continue to look forward to the occasional IACP seminar.

The time is just before the '84 Olympics and we have finalized a design and produced 10 such shields in Miller's garage, and eventually in the back of the ammunition reloading business we set up as a sidebar. We have stumbled onto a laminate and cross-stitching the Kevlar allows us to stop all handguns in both the window and the shield. Because the shield is shaped with a soft anodized face, when bullets strike it the force is distributed down rather than back.

We get an invite to display and demonstrate the shield at the San Diego police academy, a training area they call "Duffy Town." We have shot three shields to near destruction with no penetration, so Jeff takes the shield down range with the instructions to range control that he is going to lash it to a pole. I am back at the table with a variety of handguns and submachine guns ready to go. As he passes me he tells me, "I'm going to get down behind this and when I give you the signal fire as many shots as you can get off before they shut you down." Why not? What could go wrong?

I catch the Penguini after overhearing his instructions, sidling towards an escape route as Jeff gets set and finally gives me the hand signal after crouching behind the shield so he doesn't get hit in the shin by a stray round if they wrestle the gun from me. I pick up a submachine gun and empty the magazine, and manage to get a few shots off with a .44 Magnum, all while Range Control is screaming, "Cease fire!" The guns and I are semi violently separated by Range Control's safety goons. The Beast stands up and invites the attendees to view the back of the shield and himself: "Anyone want to come and take a look, no holes in it, no holes in me!" The range master is incensed but the cops like it. We stand behind our products, literally. This is not a marketing method I recommend but it works.

The approaching Olympics triggers a buying spree and we quickly run through our inventory. We are a contender under our name of Rhino Armor, so we print up glossies and start the convention circuit. This depletes our cash on hand, and it seems to be a never-ending cycle of feast or famine. In the middle of this scramble Jeff finds out he has Meniere's disease, which has him occasionally writhing on the floor with vertigo, and projectile vomiting. This can be entertaining but complicating. I am at the ammunition factory one afternoon when he says he needs to lie down. He goes outside, gets in his car and naps out. Thirty minutes later the cops show up on a complaint that a drunk is laying half out of a car in the parking lot. Sure enough it's the Beast. He is flopping around weakly and has spattered heave ammo all over the car, himself and the parking lot. Now any suggestion of caffeine and he goes into spasms. Two weeks after the parking lot incident, we are in my new jeep Cherokee returning from LA. He has an attack and is trying to get the window open while I am threatening his demise if he chucks up in the car. We get off the freeway and as we are going down the ramp, a Porsche with a sunroof passes us on the right, and stops next to us at the light. It's some LA type with the coiffed hair and open shirt with gold neck chain, and Serengeti glasses, next to his equally glamorous blonde bimbo. They both are smirking at the Beast, who looks weakly at them making moaning sounds. The guy throws some snotty remark and Jeff obliges him with everything in his digestive tract, right through the sunroof. Lovely.

* * *

Our lifestyle and the business we are in lend themselves to meeting interesting people in all walks of life. These range from the fantastic to the law-abiding, with stops along the way for the odd criminal or unsavory specimen. R. O. is one such personage. He is in the ammo business and is a gun nut/patriot supporter of anything using selective force. He's a 250-pound Bubba with one of those falsetto voices usually heard down south. Speaks Spanish and knows every cop between here and the Moon. He is an ex pro football player and moves with astonishing grace. He can also teach mantracking since he is an avid bow hunter, so we decide to add tracking to the seminar syllabus and take him along to our second seminar in Montgomery County, MD. They have a big complex with buildings for the fire academy, road track, classrooms, and an underground Murphy's town.

We are set for a day of training in rappelling and are set up on the fourth story of one of the fire buildings. R. O. is running the rappelling station, I am helping out. We have also brought Jay with us sans his little buddy Rocky. He has driven his camper down here with all the equipment. I had ridden with him. It would be a marvelous trip if you like speeding and controlled mayhem. He has the RV parked in the lot below and had retired to his cave about an hour prior. He had no more classes and soon we can hear Merle Haggard yodeling his way through the thin walls of the RV.

R. O. says that we need to set up an additional station, so I turn to one of the students. This group was ostensibly from the State Bureau of Investigation, read FBI being undercover from the locals. Nice group, well mannered, obvious newbies, getting some extra training. I tell him to go down to the RV and have Jay give him ropes, snap links, cable etc. He takes off and I return my attention to R. O. The kid comes back in record time sans equipment. I ask him where the ropes are and he gives me a nervous look and suggests I go check myself.

Jay has medical issues and carries enough pharmaceuticals to start his own Walgreen's: he has nitro tabs for his heart, and amyl nitrate snappers in case he needs a quick boost. All this is rolling through my head as I go down the stairs. The closer I get to the RV the louder the music is. He has changed to Charlie Pride, but it's at agony level. I sweep the door open and the overwhelming smell akin to that of old sweat socks wafts past me. Jay is standing in the middle of the room with a half-bottle of Jack Daniels, dressed in silk boxers, cowboy boots, and a Stetson, accented by an ankle holster with a belly gun. He pulls a baleful eye in my direction as Charlie Pride reaches crescendo, waves the bottle at me and queries if I need something. I locate the source of the smell. It comes from the crushed amyl nitrate capsules that are scattered in the carpet. He is supposed to have them on hand for his heart condition, in case he needs a jump start. There is one still stuffed up one nostril, with the string hanging down like a mini tampon. He had apparently spilled the rest when he was thrashing around, then stomped all over them. As the fumes rose the more he did his rendition of line dancing.

This is obviously the sight that had greeted the young agent. I have to give the kid credit. He didn't panic, and more importantly hadn't tried to engage the ogre in conversation. It was all my fault, as I had told Jay that we were winding down and released him to his own time. He was supposed to be doing inventory and getting the gear ready for tomorrow's training. Not expecting company, he had resorted to his normal afternoon entertainment, felt the need for the amyl boost and the rest was history.

I leave the door open and reach over into the closet and retrieve all that I had sent the kid for. I manage to get the music off because the master power switch is next to the door. As I start out the door he lurches toward me, mumbling about all the intrusions.

I explain the situation to him in brief, not wanting to hear his side of the incident. I step down onto the pavement and before I close the door I tell him to put some pants on, lock the door and don't open it until I knock three times in sequence.

I get back up to the top and start setting up the second anchor point. Jeff wanders over to tell me the Feebs have been acting strange. I give him a quick rundown on what had transpired. He keeps head-jerking through the explanation and looking more and more unsettled. He mentions that he hopes Jay stays out of sight which is pretty much the nexus of shared caution so the situation doesn't get out of hand. The kid who had witnessed the ogre gives me a nervous smile. I point at the RV and make a hand motion imitating the dangling amyl nitrate capsule in the nostril and mouth "heart condition." It's weak but at least it's a possibility.

We get done, dismiss the students, and break down the station. I let Miller go ahead while I watch from the roof as he moves purposely towards the RV. He bangs on the door, no response; bangs on it again, still no response. He bangs on it a third time and the door swings open, Jay looks the same except he has pants on, he reaches down and drags Jeff inside and slams the door. It looks like one of those National Geographic films of some tropical arachnid ambushing its meal and pulling it back into its lair. I need a beer.

* * *

We get rave reviews on our seminars, so much so that we fill up a few sessions which we use as a marketing tool to sell our courses and the new SWAT shield. We start to include it in our room clearing and barricaded subject classes and practical exercises.

We have a gig in Tucson, Arizona. The same drill but we have become more refined. It's the same crew except we have replaced Jay with a homicide detective who runs the SWAT element in Costa Mesa, California. We have also added Steve, who I had known in Berlin and was on the assault team with me. Steve is an imposing six-foot-something bundle of why the Irish are part of every melee of note. He is intense to say the least. R. O. has acquired a class A motor home and we had all driven over to Tucson from Orange County, California. It's six hours on the road, which went by in a blurry recollection of frightened faces of other motorists, and maneuvers that defy sense. Looking like a Viking raiding party we arrive in Tucson. We are greeted by our hosts who include Linda Ronstadt's brother who is the chief of police. Miller has some history with the Ronstadt family as he met Linda when she sang at the Troubadour on Santa Monica Boulevard in Hollywood and he was working undercover for the Bureau of Narcotics and Dangerous Drugs.

There are several departments from across the states, and a foreign student from Taiwan. The gent from Taiwan is a smallish but wiry fellow and very aggressive. We are running a dual station with room and stairwell clearing, and rappelling in the same building. We are using air soft and paintball guns against aggressors made up of a third set of students. We rotate everyone every hour, and they move on to the next station switching with the team on that station.

The Taiwanese guy has the shield and refuses to give it up. He has to run up four stories with the shield as point man protecting the rest of the team. He does it several times before Miller takes it away from him. I am exhausted from just watching him.

Halfway through the exercise R. O., who is running the rappelling station, reaches down and lifts up a sheet of plywood on the roof, asking what it might be covering. No one is looking at that moment, but they hear a muffled cry, but when they look over R. O. is nowhere to be seen. I lift the plywood again with Jeff's help. Turns out it was covering a four-foot square hole in the roof. Looking down through the hole we espy R. O. sprawled on the floor and weakly moving. We rush downstairs and quickly give him the Special Forces medical evaluation. He's breathing, bewildered and beaten up, but still moving so he is okay. Any complications will be dealt with later. Thank God for his pro career, it was probably reminiscent of being run over by Bubba Smith.

The second night in Tucson we are in the RV cruising the Miracle Mile, the "unofficial red-light district of Tucson." There are plenty of soiled doves along the route and we are giving a rousing evaluation of each as we cruise by. Steve and one other are in the mood. So we pick up two of the ladies and are directed to the "by the hour" hotel where they ply their trade. The loving couples leave the RV and head upstairs.

We can see the second-floor walkway and both couples go to separate rooms. I am sitting there with Jeff and Bill the homicide dick, discussing venereal diseases and the chances of both of our comrades developing symptoms after we get back. We are having a great time, sipping whiskey and regurgitating the symptoms of the worst cases we have seen so far in our lives, when not five minutes later the lady who had gone with Steve jumps out of the window of the room, half dressed, and runs off into the night. Bill being a cop assumes that she has rolled Steve and that we will find his unconscious body sans wallet if we go up there.

We start to do just that when Steve opens the door and saunters out and down the stairs, met by the other night reveler. They both get in the RV. We ask them how it was, they both chirp in "Best I ever had." We know that in one case that isn't true. We start asking Steve why the woman bailed out the window but he goes Buddhist, silent and staring off into the distance. Bill deftly wheels out onto the Miracle Mile and we espy her at her old spot adjusting her outfit. Bill practically runs the poor woman down with the RV, stops and demands answers from her in his best cop fashion. She is in a nearly hysterical rage, screaming and pointing at Steve, "Keep that crazy mother-fucker away from me, I wouldn't do that if he had all the money in the world!" She skitters off into the night and all our attention focuses immediately on Steve, but we never learn what the thing that he wanted was.

* * *

All in all, our dream of a training and equipment empire is a bust. There never is enough money to properly expand or to market effectively. We continue to squeak along looking for investors and ending up with either people that have the money but you won't take it, or shysters that want to steal your ideas and do their own thing. Not having an MBA degree but with a keen bullshit filter we manage to avoid the outright mugging, but our filter is flawed.

Our journey has only begun; we are doing well enough but not getting rich by any means. We have discovered that we can actually survive outside the great green womb of the military. The cautionary here is that any chance at an adrenalin rush always makes your ears perk up. When you have run with wolves it is hard to like puppy chow.

CHAPTER TWO

There are Carpetbaggers in Our Soup

We have met the enemy, and it is us.

—WALT KELLY, *POGO*

Sometimes learning the hard way is the only route to your objectives. We have learned from our mistakes to a degree, but every day brings new challenges. Being a small business is supposedly the backbone of America. We can say that the lumps one takes because of flawed data and expecting those in businesses to have the same ethos as you, could break the strongest of wills. But no brains, no headaches is a good excuse.

We are constantly looking for new ideas for equipment, but our sales of the ballistic shield have slumped. We don't have an intact budget for marketing and production, so the shields, while hand crafted, aren't able to make a dent in the market. There are a number of ballistic innovations that resemble ours and one unique item, the ballistic face mask. I don't know who came up with that, but a mask designed to stop penetration by a .44 Magnum only results in a better-looking corpse. The blunt trauma produced by the round striking the mask would be sufficient to smash through elephant physiology. The human cervical is not designed for that kind of force and would snap like a dry twig.

When we had begun the development of the shield, we had gone to one of the Penguini's friends who was a manufacturer. He assisted in the design and initially produced the prototypes and production models. He, being astute enough to see the market, and knowing we didn't have the money for patenting or copywriting, soon went into business for himself and was ultimately responsible for the shield being in the hands of police departments across the US and eventually highlighted on every cop show from *Blue Bloods* to *SWAT*. No patent or copyright, we have no claim of infringement. We were soon pushed out of the market and were selling our left-over inventory. That whole crowd was from Pittsfield, Mass. which is like crime central on a normal day, akin to a riverboat encampment along the Mississippi circa early 1800s.

Jeff has wrangled a contract to do an independent survey of the security situation at the 1984 LA Olympics through Edwin Meese. We are the perfect pick. This is a classic Red Flag operation. Look at the whole picture, break it down into tasks and find the weak points. It works out well—we are able to get their focus on the areas where we believe they have problems. We also manage to sell a few of our shields to various departments. Years later Jeff will meet Mr. Meese at the Heritage Foundation in Washington, DC and he will thank Jeff warmly for the detailed report submitted. It had resulted in bringing the brass from both LAPD and the FBI to a meeting at the White House to iron out their differences. Along the way Jeff is introduced to David who is the security chief for the Israeli team. Once the Games are over David decides to stay in the States and gets a gig as bodyguard to Elizabeth Taylor. David is an Israeli who parlays the country's expertise in security into a thriving business protecting "the stars." Ms. Taylor is his premier client. He soon has work for us in his growing angst empire.

One of the first jobs that he gives us a role in is the security of the mistress of a swain that manufactures toys. They had hired some retired LAPD type, who after being terminated for attitude had now taken to stalking the lady. David always has one guy on the site to watch the gate and walk the grounds. The pay is good. Good enough that I am driving two hours each way every day. The LAPD swiffle soon gets the message and moves off so it's a nice day in very pleasant settings. The guard station is at the front gate, with the switch in the gatehouse and the grounds beyond. There is a converted carriage house that now serves as a garage, with French windows facing the inner courtyard. Not a dangerous gig except for the fact that the lady is a complete dingbat. Great to look at, but the elevator doesn't go to the top floor. She has two Rottweilers who are trained attack dogs that stay by her side constantly. She always forgets to call the gatehouse when the dogs are outside. So far, they have bitten David and everyone else, but not me.

I arrive for my shift and get set up. I had brought along a sack lunch of lobster sandwiches and some coleslaw for the day. I place it on the folding chair and go into the garage. I turn the alarm switch back and turn to go back out. I never get to opening the door because the two Rottweilers are not only out, they are busily devouring my lunch, bag and all. I am incensed. She had let the dogs out without phoning the guard station.

The dogs finish my meal and approach the French windows with hostile intent. I crack the door, and both start growling and ease up to the door. The only thing I can think of is both a statement of my anger and disdain for their little canine bullshit, so I piss on both of them through the gap. This has immediate reaction from them. They look at me, spin around a couple of times, sniff each other, then look back up to me, then their little tails begin wagging. Aha! This is pure alpha male stuff. Over the next few hours we become friends. One is laying at my feet and the other one thinks it is a lap dog. The bimbo never comes to check where her pooches are; finally David shows up because apparently she actually had been calling for the dogs, and had also called the gatehouse (I refused to answer). He looks through the spyhole, sees me and the dawgs, assumes that I have somehow tamed the beasts, then opens the sally gate to come in. The dogs immediately go for him. He barely escapes with a torn pant leg. From then on she would let them out and they would make a beeline to see if their new pal was there with treats and rough-housing.

One of our tasks is to accompany her when she leaves the *casa del amour*. Unless you have pulled this type of work before; accompanying a woman on her daily rounds is the most mind-numbing, tedious, boring task next to watching golf and the shopping channel on sedatives. The secondary mission is to report on any indiscretions she had with any male along the way. I don't care to be anyone's snitch, but the fact of the matter was, she was having sex with her art instructor, her hair stylists, her mechanic, and several others. I tell David, but I am sure he never passed it on to the client. We just try to keep a lid on it. It is an interesting gig.

That dries up but a month or so later he calls us back and asks if we can help with securing a site where Liz is premiering her new perfume. It takes place at the home of the guy who owns Herbalife. We get there and park the car on the adjoining street and go inside. This place is a true mansion, elegant, spacious and soon to be full of the Hollywood elite. We are given our briefing and Jeff and I are assigned a roving supervisory role. Most of the guys are Israelis, all claim to be former Mossad, which is bull-puckey, but they are conscientious and do their jobs. We station ourselves on the lower level down from the main event where there is a fountain and courtyard outside the lower level, which contains a mini casino with slot machines and two enormous bathrooms. These are reserved for Ms. Taylor. This is where we station ourselves when we aren't roving. Roving consists of checking on the security guys and being innocuous. There are lots of sights to see. There are stars everywhere. One of the highlights is Peter Falk, who shows up with a woman who is absolutely beautiful with a dress so low cut that it defies gravity and seems to be held up by her nipples. Falk tells us as an aside that he could rob every male here and they would never notice. The other highlight is Zsa Zsa Gabor, who is half tanked and takes a shine to Miller. She attaches herself to him and is calling him her Sweet Captain Nemo. Miller bolts for the men's room anytime she passes near as his only defense. Disappearing from the crowd like his new namesake. I rename the men's room the Nautilus.

Lots of stars, lots of their hangers-on. Some know about the facilities and keep coming down to use the restrooms; we dutifully turn them away and explain that they are reserved for Ms. Taylor. Most are gracious though disappointed. There is always one who thinks they can bull their way, and one such Twinkie is a star in the hirsute world of styling. He wears this short little semi-Panama and has a long braided pigtail that reaches to mid-back. He starts through the door, but Jeff stops him and the little weasel gets all huffy, demanding to know if we know who he is, yada, yada, yada. We tell him politely that he has to use the facilities above, but he isn't taking the message. He tries to push past Miller who stiff arms him and he cartwheels into the fountain, splashing down in the lily pads and imported frogs. When he surfaces, he has lost the hat and apparently the faux pigtail as well, and emerges bald as a cue ball. He rushes off with threats that we will never work Hollyweird again. Good riddance, little twit.

The night wears down and Ms. Taylor arrives for her departure which is also through the casino, and out to where her car and the escort vehicles are. Jeff and I are trailing her crew by a few feet as we enter the casino; we take a right, past the slot machines when up pops this skinny little swish clad only in his birthday suit and a pair of black dress socks. He lunges for Ms. Taylor, while gushing how he had admired her from afar, had watched all her movies at least 10 times and had dreamed of meeting her. She handles the scene with the grace of a queen, but he is swarmed by her praetorian guard like an Israeli armored column. It is a high point of the night but still doesn't compare to Falk's date.

* * *

Shortly afterwards we get a gig to explore the possibility of presenting our programs and products to the US government at the national level. With the new angst that terrorism presents, we feel that we can find a niche. Jeff has arranged a series of meetings starting with the Department of State, the Pentagon, and the nuclear elves.

We arrive, and are met by Max, our ghostlike spook asset. He is dressed for the occasion in the standard DC uniform, soiled, wrinkled blue suit with a nice silk tie, this one with a very fetching mustard stain. He is wearing the obligatory London Fog trench coat, flapping in the breeze. He has his VW bug as our transportation around town, explaining that it's easier to find parking spots. Later we surmise that the sidewalks are his favorite spots. He merely tears up the tickets. Apparently, he had become such a management nightmare that the agency had loaned him (or given him, or exiled him) to the Marshal's service, also apparently with the authority to enforce his own peculiar take on infractions of law and order. This connection will result in other problems later during another IACP seminar. But for now he is our tour guide of the halls of government.

Our first stop is at the Department of State, office of their counterterrorism guru, Lou. Now Lou is not the most obnoxious specimen of the herds of bureaucrats who wander the halls and offices of DC but he does have that polished arrogance that the Foreign Service and its minions affect. We are ushered into a conference room with his nibs seated at one end with Jeff and I on either side of him, Max takes a seat at the opposite end and leans back in his chair looking all the more like some cheap rendition of Broderick Crawford. As we begin our pitch, Lou listens, digests our suggestions, sits up, straightens his tie with a manicured paw, and begins to tell us why we are all wrong and that State still has the helm on the terrorist problem. His arguments are pure bullshit, because we are intimate with the same sources he has. These same folks had painted an entirely different picture to us, bordering on panic and chaos.

Lou has been droning on about how important it is that the matrix fits the solution, legal and international consequences of abrupt and hasty actions, and a host of other government minefields that only the Foreign Service can address. Max all the while has been slowly cooking at the other end of the table, and making snide remarks that Jeff and I can hear, but Lou cannot. It's some weird ventriloquist twist they must teach at the deceit and betrayal course. I am biting my lip because it is hilarious: Lou says something inane, and Max assures him that indeed he is the expert and that despite his having no grasp of the situation, he is assuredly the top of the class in expertise. Mouthing his comments just below the level that Lou can hear, but we can.

Yes, Lou, you are the biggest windbag in Washington, oh my God Lou, thank our luck we have you to guide us, all in sotto voice. Then Lou stumbles out of his box. He points out that he has no specific knowledge in a certain area and recommends that we go ask those loose cannon idiots at Langley. He should have done his homework. Now Max is raising the comment volume and Lou finally asks him if he has something to add. Max replies, no, that he had confirmed his suspicions and had verified his assessment. Lou never gets it, he thanks us for making the long trip, but as you can see State has everything under control. This is the government version of a hand job. He ushers us out and we gather ourselves for the next meeting at the Atomic Energy labyrinth. I notice that Max has more paperwork in his hand than the leather folder he had entered with. As we go down to get the car he is full of invectives for our new pal Lou, just short of asking us if we could help him dispose of the body.

We wind our way over to the nuclear folks' hive and are ushered up to see our next appointment, who is even more condescending and arrogant than our new pal Lou. The receptionist's disdain at us for penetrating her lair is evident from the very beginning. "And you are?" she queries in a supercilious lisp while looking down her nose through designer glasses. "Mr. H.," replies Max. Letting out a sigh of boredom she peruses her appointment calendar: "I don't see your name on the calendar." "Well just call, I'm sure they will want to see us." We are immediately shown into the inner sanctum where we go through the same drill with the same results we had at State. On top of that Max is sitting with one leg crossed over the other in full-on slouched man spread. I look down and realize that I can see the bottom of his shoe, which is pointing right at me. There is a significant hole in it through which I can read yesterday's *Washington Post* article that he had patched it with. Max is done being subtle, on the way out he openly insults the grande dame of the outer office and she all but throws us out of the building.

As we are leaving Max picks up his notebook that had been perched on her classified in-box, palming the contents of the box as he does so. As we are leaving, he transfers those into his notebook, deftly gloms a fire extinguisher off the wall and hides it under his trench coat. As we are going down the elevator he engages the capitol security guard in conversation and heists his badge and laminated ID. This boy is a walking klepto tornado. We get in his VW and he tosses his keys to Miller, telling him to drive. As we are wheeling our way back to the hotel he is pulling documents that he lifted from the snotty lady at the nuclear office. Most are classified, a few are not. Max has a running dialogue about officious, pandering federal drones, as he gathers up the papers, lifts the cover sheets to briefly scan them and then flips them out the window of the moving car onto Massachusetts Avenue, right in the heart of Embassy Row. "Let's see that bitch try to explain this," he shouts triumphantly.

We swerve to avoid a minivan with the livery of a local florist on it. There is a small bump and we and the van driver stop to assess the damage. The driver is a slight Arab who is excitedly pouring forth his frustration and indignation. Max doesn't hesitate. He whips out a badge and demands his driver's license and registration. Then he rips them up and beckons over one of those uniformed diplomatic security twinkies who's patrolling Embassy Row on his little moped. Max accuses the Arab of running into us, again flashes the badge and leaves the poor sod there with no papers to the whims of the dip-security who obviously wants no part of this. We roar off to the safety of the hotel. Perhaps we can use his talents to get out of paying for the drinks at the hotel bar.

While we are in DC the Vietnam Veterans Memorial is scheduled to be dedicated by President Ronald Reagan. The night before, one of our Special Forces compadres shows up, in the form of Sergeant-Major, General, Doctor, Professor Alan F. Al served in MACV-SOG, got out and went in the reserves while attending Tufts and Harvard for a fistful of degrees, retired from the SF reserves as a sergeant-major and was immediately made a colonel by the governor of Virginia so he could take over the Foreign Languages Department at VMI. He was later promoted to brigadier-general so he could become a dean. The alumni avoid him like a leper and don't consider him a "real" general. He is still an SF NCO at heart, which doesn't endear him to the hallowed halls crowd. We are accompanied by an FBI senior something-or-other from Quantico who is interested in our ballistic shield.

At the base of the wall visitors have left many offerings, one of the most popular of which is pairs of jungle boots. While we are scanning the wall looking for the names of friends, Al sits down and starts trying on the boots. He finds about four pairs that fit and ties them to his belt by the laces. As we leave the site our FBI friend is shocked and appalled by this behavior but Al is totally cool. "Don't I deserve them more than some Parks Department twink?" Flawless logic as far as Miller and I are concerned,

Our mission to DC is a complete bust, no one wants to address the terrorist threat, no one even wants to admit there is a problem. They had gotten through the Olympics, so it wasn't on their radar.

Less than two weeks later we are back in Gaithersburg for another SWAT seminar. We have Bill the detective from Costa Mesa as one of the instructors. Bill is always a joy to work with, his favorite retort is that he has a nice ensemble in the form of an orange jump suit that he can arrange for us. He is a great instructor, though, and funny. We have also invited Max back for another stab at being productive, demonstrating a masochistic tendency that will haunt us the rest of our lives.

As usual he is late. Jeff and I are in our hotel room going over the next day's syllabus. The phone rings and I answer it. It's Bill and he tells me that someone is at his door and he thinks it might be one of our friends. I ask him how he came to that conclusion. He merely says look out your window at my door. We both peek out the door and there is Max standing next to Bill's door like he is front man on a room entry. He has on a black leather jacket and appears to be wearing sap gloves and has a pistol in his other hand. This is not what we need to end the day. I lean out and *psssst* him over and we drag him into our room. He is all chuckles about the neat practical joke he was about to pull.

He eventually gets around to why he is late and explains that he can't work with us this week because he has a new gig with the US Marshals. Why not? He has the Baby Huey physique which hides the fact that he is a brawler and likes the pain. It seems that the government, mindful of the fact that he has been in their service for enough time to qualify for a pension eventually, finally found a slot that fit him to a tee. He is on some "dangerous felon" hit squad that only goes after the ones that are rabid. We had anticipated his absence, because Max is after all Max. We move the schedule around so that we can cover his classes. I am ever the observer and I notice he keeps eyeballing the keys to our rent-a-car, which is a brand-new Firebird, complete with the Burt Reynolds sunroof and massive decal on its ebony hood.

When we had rented it, I had gotten the full-on, walk-away insurance policy, mainly because we had the funds and I know all our driving habits. The girl working the Hertz desk had cocked an eyebrow at me and cautioned me about accidents as this was a new addition to their fleet. We had, of course, immediately taken it to the track at the police academy and drove it like we stole it. Now Max has some sort of scheme to separate us from it for his own prurient interests. He comes up with some bullshit story that he has to leave his car because he doesn't want it to be identified and that he just has a few errands to run. If anything happens, he assures us that the US Marshals' service will be responsible etc.

We relent and give him the keys. He is supposed to be back by midnight, but typical Max, he still isn't there until near noon the next day. We are in the underground Murphy's town when he breezes in like a waft from the sewer. He is moving in this jerky motion that he affects when he has bad news. He tells us that he had a minor fender bender the night before and has spent the day taking care of it. We ask how bad the damage was, and he says we should come outside. We walk out into the sunlight and as our eyes adjust, we see our Firebird loaded on the back of a two-axle trailer, and it is twisted wreckage, looking like some modern art depiction of a crushed pretzel. There is a government pickup with park service markings towing the wreckage. The front end of the Firebird is barely recognizable.

Apparently, Max had been running his errands when he spied one of the felons on his wanted list. He had chased the guy down through alleys and side streets, run him off the road then backed out and as the culprit started to make a run for it, he had crushed him against a pole. The car was totaled.

We have to go with him to where we rented it. As we pull up, I can see the girl who suspected that we would bring it back damaged.

She shoots out the door before the trailer comes to a complete halt. She is shocked and angry and I am the root of her focus. "I knew you would wreck this, I knew that you weren't trustworthy. Look! Just look at this!" I'm thinking that it is a good thing she hasn't gotten word of our antics on the track because she would probably spontaneously combust. She is acting like it was her car.

Max of course whips out his badge and starts to explain that the Marshals will pay anything not covered by the insurance and that it was damaged during a "law enforcement action." She is not having it, not at all, actually tells him to shut up, she is talking to us, the responsible parties. It goes back and forth until Jeff and I leave them with the keys and the express checkout envelope and go back to the academy. It eventually blows over but we are on the "do not rent" watch list at Hertz for a year at least.

Max blows out of our lives like an ill wind. We find out later he had come up with a novel approach to luring felons. They would find where the wanted fugitive was hiding, then inform his relatives that Max was from the Lottery, or Disneyland, or Princess Cruises, and that the perp in question had won money and/or a free cruise to the Caribbean for six people and that they would send a limo to pick him up. Of course, on the eventful day the felon, family members and one or two accomplices would be at the pickup address when the long, cocaine-white limo pulled up front. To the cheers of the crowd the liveried driver would hold the door for the winner to get in, open it and as he slid in four marshals in the back would gorilla gang him, handcuff him and read him his rights. Sometimes they would get more than one on the same grab.

We heard through the mojo wire a couple of years later that the Marshals had traded him to the Border Patrol or Bureau of Land Management or some other innocuous federal agency. Too bad they don't have rookie cards; his would have been worth a fortune to the collector. We ran into a couple of grizzled marshals at some convention a few years later, and asked them if they had ever heard of Max. We got the same reaction from both: they rolled their eyes and excused themselves. Hopefully he has gone back to school and gotten a teaching degree. I bet that would shake up the liberal pipe-smoking socialists in the education cabal.

* * *

We have moved into a new office since the ammo plant and IACP contracts stopped coming. But we still have the shield and thanks to R.O. we have a whole line of specialty footwear made for us by VANs. We have also taken on a new partner who came out of the blue. He came in with all the bluster and dialogue of a business guru with the genuine objective of turning our efforts into a real business. He offered to roll our business into his infrastructure and in return we would get ownership in the joint venture in the form of stock. We have also opened a relationship with a small firm to produce a line of lasers that fit on pistols and rifles. These are primitive compared to the digital devices today but cutting edge for that time.

Our new partner is from Vegas, and acts, looks, and presents himself as an entrepreneurial genius. He moves our operation to a set of offices in Fountain Valley and we set up shop. The Badger is a six-foot-plus blustering dialogue of the great things that he can accomplish with us as the raw material. Loud mouthed, with an equally loud wardrobe, he at first adds the bit of zip that we need to make things work. He takes the executive office, and we each get our own office and we have a conference room. We also now have a secretary-receptionist, who turns out to be the jewel in the mix.

Almost immediately our newfound benefactor appears to have some bad habits. It starts with the morning meetings, which are daily unless you can slip out before his eminence rolls into the office. These are long droning diatribes about how there is no ego in business. Translated, there is no ego except his, and it is the size of a house. He takes possession and control of the books and checkbook. Not too long after we start getting angry calls from our vendors about not being paid. The Badger covers his heinie by paying the most in arrears but it's only a harbinger of the problems to come. He also buys new furniture for the office, most of which goes into his lair. Additionally, his wife, who fancies herself as an interior designer, orders 1,700 dollars of plants for him for the same. It looks like a bloody jungle in there. Our protestations are to no avail, he has control of the company.

We have a number of firearms that are on loan from the manufacturers to use in our training and market promotions. We find a couple missing and when queried, he explains that they were part of a deal that he made to borrow money for the company using our hard assets as collateral. Two shields are also missing and the rest of the inventory has tags with numbers attached. We begin to gather our wits and try to keep the small fires from becoming a conflagration.

The only bright spot is our new office manager, Vanessa. Vanessa is a mere sprite of a woman with a sardonic sense of humor and she loves us, all except for the Badger, whom she loathes. He has her fetching his dry cleaning and other mundane chores until the Penguini and Jeff corner him and tell him to do his own fetching. This sends the Badger into a week-long pout about how unappreciated his efforts are to save us from ourselves and sloppy accounting. I am at the point where I just want to pick a spot to leave the body. But our interceding on her behalf cements Vanessa's loyalty to us. In return she begins a campaign of misdirecting things that are blatantly wrong and conducting a guerilla war against the giant ego. He can't fire her because we need to agree. He also tries to jerk her off for her pay until the Penguini goes in and tells him that he is going to pull his arms off and use them as a switch on his genitals.

Vanessa speaks four languages fluently and has an MBA. She is the reason that we even survive this period with some shred of what we started out with. As this soap opera unfolds, we still have the touch at attracting the lunatic fringe. Almost on cue, a nefarious personality named Louis shows up through an outside contact. He wants to buy, of all things, weapons, lots of weapons! We advise him that a valid end user certificate is required. He assures us that will be no problem and that the French are somehow involved. This is malarkey on the face of it. If the French were involved, he wouldn't be in the picture. He says that he will explain in depth when we meet and wants to speak with us about a training advisory program. We agree to see him and on the day before Thanksgiving he arrives for his appointment. The minute he walks in he is greeted by Vanessa, who asks him to have a seat. He doesn't sit long: he is nervously pacing and occasionally peeping out the shades. Vanessa calls us on the intercom and suggests that we come out, adding sardonically that our guest is exhibiting all the signs of someone on the lam.

We watch him for a while before asking him to come to the conference room. Louis is a gaunt, stringy, boiling vat of suppressed energy, of French descent. He had served time in the Foreign Legion. Now he is acting as some sort of military advisor to some deposed African despot, who wants to take his homeland back from a tyrant, a very common theme in those days. In this case the rightful president is apparently living in Brooklyn, and according to Louis, his group has the complete backing of the CIA for their coup (sure they do!). Louis lays out this plan that is like a carbon copy of the film *Dogs of War*. He also has a shopping list. He tells us that he has upfront earnest money. We ask him how much and he says two and a half grand. Moreover, it is in a brown paper bag duct taped to his chest. He retrieves it in front of us and seems to enjoy the tape being pulled off. We tell him that he is going to need a lot more money than this to fill the shopping list that he has. He agrees to speak with his principals and leaves the office, skitters down the walkway and disappears in the foot traffic.

By now all our alarm bells are ringing so we call the Old Man, Miller's buddy who is pretty high up in the CIA pecking order. He listens to us then pauses for a moment as he makes a few phone calls. He comes back on and asks us if we have any friends in the FBI; we respond "Sure." He suggests that we should get in touch with them; we agree to call them next Monday. He tells us no, and to call them now, because the FBI is already in our parking lot. So it's like that is it?

Sure enough two polyester twins exit a car and make their way up to our office. We get the finger-wave briefing of how they have been following Louis and his client, who is an exile living in public housing in Brooklyn using the diaspora from his home country as a donation base to back his invasion plan. We never hear from Louis again, and eventually he and his FBI tail disappear into the woodwork. We assume that it is all over. We are the good guys, have duly reported the incident, and use the two and a half grand to do repairs on one of the reloading machines. But the whole scheme will pop up like a bad penny for years to come.

Vanessa, bless her petite heart, has the Badger in a complete state of frantic scrambling. She hides receipts from him, keeps a secret watch on our bank account, and we begin to notice strange expenditures. Vanessa enlightens us and says that she suspects that he is rolling debt from his previous companies and using us as his new vehicle. She can't say his name without spinning around three times and spitting.

Our new product is being developed by a company in Laguna, Trac-Tel, headed by a brilliant innovator with a staff of equal geniuses who make up the lab. Great staff, impressive designs, they churn out camera devices that can be used to view under vehicles, remote sensors, and of course weapon lasers.

Jeff and I are going to Korea and meeting with the chief of procurement. I had known the good general when he was a lieutenant with the Tiger Division in Vietnam. He is now a major-general and has scheduled our meetings to be on the island of Cheju Do, off the southern tip of the peninsula, where the army has a rest and recuperation facility. This turns out to be a 700-acre first-class resort, complete with casino, spa, and 36-hole golf course. The only drag is that we have the design company's CEO with us. He is a droll diminutive Brit who is a pain in the ass and has arachnophobia. Fred, the brains behind Trac-Tel, had picked him up from being the CEO of a firm that produced fine china dinnerware. It must have been a fire sale as far as we are concerned. He spends most of his time rubbing elbows with anyone that seems important, pontificating obscure business formulas, and the casino. The general can't stand him and calls him the British poofter, but despite this our meetings result in their willingness to buy a couple of thousand lasers made to fit the K4, which is the Korean copy of the Israeli Galil.

The Badger is soon in high-speed coverup. We return from Korea, where we had put expenses on our credit cards to the tune of ten grand. Normally I would file an expense report, but Vanessa calls me aside and tells me that there is no money in the operations account. I don't understand because 10 days prior there had been over

50 grand in it. She shows me the account information. Badger had withdrawn all but about 15,000 the day we had left for Korea. Badger is gone for the day down at the local bistro holding court. Miller and I toss the executive office and find that the closet that held all the weapons is also bare and some other equipment is also missing. Penguini breaks into the Ego's files and desk and we all begin to go over the books.

By morning we have discovered that the Badger uses up companies like changing diapers. He strips the assets, rolls the last victim's debt into the new one and creates yet a third with all the assets. Vanessa comes in at 7a.m. and we tell her that she probably won't have a job by day's end but we will make sure she is current and gets two months' severance pay. She wants to stay to see the burning at the stake, but we convince her to take a long breakfast.

The Ego arrives at 9a.m. and we are in the conference room. His office looks like two Marine drill sergeants have ransacked it. He lets out a little yelp and is midway to getting on full bluster when we rush him into the room where he cowers behind his desk as we lay out the evidence in front of him. Penguini and I tip the desk on top of him and are trying to strangle him with the back edge when Jeff pulls us off. We make him write Vanessa's check and another to cover half of the expenses, which will empty the account. We leave him with a warning that we will skin his ass if either check bounces. Vanessa comes along as we are leaving and we give her the check and tell her to cash it today.

Badger flees town with the speed of Zorro, stripping the furniture and plants and anything sellable, and runs off to Vegas. He also leaves a three-grand tab at our favorite bar. We pay half of it so we are square with the owners who are friends. We are soon out of work and open to suggestions.

CHAPTER THREE

Looking for High Tech in All the Low Places

We have fallen from a somewhat steady income and we are looking for salvation, which comes along in the form of the company that was developing our concepts like lasers and camera applications. Fred is an amazing technology innovator who had helped invent the voice stress analyzer, made famous in the movie *The Trial of Billy Jack*. This device was slowly replacing or enhancing the use of the polygraph.

His company Trac-Tel is staffed by some very talented and genius-level research and development guys, with their own clean room/lab where they are sitting in an environment of radio frequency energy that probably isn't healthy. A casual observation finds that they are either bald or have hair like Garfunkel. They turn out some amazing work though. We avoid going into the lab at all if possible. Fred has located the company in Laguna Canyon because it was RF (radio frequency radiation) free. It's a beautiful setting with Laguna just five minutes down the road. The only suck in the deal is that it is an hour's drive to our homes without traffic, up to two hours during rush hour which in southern California can be six or seven hours a day.

Fred saw some glimmer of what he hired us to do which was help design tools using his technologies, and act as the marketing arm to government. Fred and all the geniuses are upstairs with the Brit and the lab. We are on the entry floor along with the receptionist/office manager. Christine is smart, sarcastic, quick-witted, gorgeous, and an aerobics instructor in her spare time. She luckily takes a shine to us, so we have a symbiotic relationship. She covers our ass and we serve as constant entertainment.

We have the laser and we are developing camera applications, all keyed on either law enforcement or military use. We have a variety of weapons in our office safe fitted with the laser, and are at the range usually twice a week perfecting them. Less than a month after Badger's disappearance, Jeff and I are in our office cleaning equipment and guns, having just come back from the range. We get a buzz from Christine saying there are some unpleasant men in the foyer asking for us by name. We tell her to show them in.

Bubba, the one in the front, is a very big man but going to seed. His partner Guido is a compact-looking swarthy individual. After Christine goes back to her desk, we ask them what they want. Bubba says he works for B. C. and that we owe him money.

We knew that the Badger had made a number of deals and we had suspected he had borrowed money from questionable sources. B. C. was a loan shark with reported mob connections, and we had met him socially once. After that the Badger had become his bosom buddy. It turns out that the Badger had used all the company guns as collateral on a loan from this shark. Since he has fled the state, the shark thinks we are now on the hook for his debt. Jeff goes into a protracted explanation of how that isn't going to happen along with a message to take back to their boss about biting off more than he can chew.

Bubba the giant leans on my desk as Guido gets all jittery. Bubba looks at me and asks if I have anything to add. I reach over behind me and pull up the fully loaded, laser equipped submachine gun, which I had been planning on cleaning next. It still smells like burnt cordite. I point it right at his gut and tell him that I am curious as to how he and his little friend expect to get out of the building on their own, and pull the cocking handle back for punctuation. His eyes and the little weasel's go wide and they boil out the office door, past a startled Christine and out into the parking lot, get in a white caddy and peel out, heading east on Laguna Canyon Road. They almost hit two cars in their haste. Jeff tells Christine to call the police and we give them the details of the confrontation and the description of the two.

The Laguna PD know us. We swap equipment and sometimes training with them so they are more than helpful. They arrest the two and take our statements. We don't hear squat until the shark calls me in the middle of the night a few weeks later, to say that someone shot at him. I tell him it couldn't have been me because he is making the call. We upgrade our situational awareness in case they really are that stupid. Later that whole crowd is involved in a case with the owner of a strip bar (that mysteriously burns down) and his girlfriend being murdered and the shark taking yet another .22 caliber to the brain housing group and being blinded. We also find out later that Bubba, who the shark used as his muscle and bodyguard, turned up dead in a parking lot with two .22 rounds in the head. On top of that it was revealed he was in the witness protection program all the time he worked for the shark. Badger is still on the run but we wash our hands of the whole affair and get back to business. We had hoped our cool-headed handling of the situation would at least impress Christine, but she continues to treat us as a wrinkle in life's leotards.

We decide we need a short break and grab our friend R. O. and head out to Ensenada for a long weekend. We drive down from Orange County, check into the Bahia Hotel and immediately head the four blocks up Avenue Lopez Mateos to Hussong's. For the uninitiated, Hussong's is the closest thing to a Wild West saloon that can still be found anywhere. We start with beer, but soon switch to margaritas and by the time the cruise ship passengers flood the place we are well lubricated. R. O. finds a schoolteacher he is trying to put the moves on, and prevails on me and Miller to entertain her 13-year-old son. Just what we need on a bender in Mexico, a babysitting gig. But there is a code among guys, so we say sure, we'll keep the kid entertained, take your best shot. I am interrogating the kid about his life and experiences and not too surprisingly discover he is still a virgin. So of course it gives me a mission. I leave the kid uncomfortably in the grip of the Beast and go trolling in the target-rich environment of Hussong's. It doesn't take long to find a drunk college girl in a Papa's and Beer T-shirt who is enthusiastic about the idea of breaking the kid's cherry. I drag her back to the table where the kid is sitting back to the wall, wide-eyed, while the Beast regales him with stories of mayhem and madness. I turn the girl loose on him, and she does her best trying to seduce him, but by this time the kid is terrified and having none of it.

R. O. shows back up with mom and the kid can hardly wait to spill the beans about what has been going on in her absence. She immediately turns her fury on R. O., raging about his horrible, immoral friends, and stomps out, kid in tow. Oh well, the best laid plans and all that. Nothing left to do but order another pitcher of margaritas.

Somehow that night we manage to find our way back to the hotel. I hear stories later about crawling in the gutter, biting the front tire of a cop car, stealing tacos from a roadside stand and hanging on the front door of the hotel while other guests went in and out, but I can't really attest to the truth of these rumors. I do know that the next morning I have a nearly terminal hangover and R. O. and the Beast look nearly as bad as I feel.

We find a little café where big platters of *huevos con chorizo* and a lot of coffee bring us nearly back to the brink of humanity and go looking for a place to enjoy the day. We wander into an outdoor cantina with a flagstone floor and metal patio-style tables. We pick out a round, umbrella-ed table in the corner and start ordering Tecate beer, which comes in cans that we can build a pyramid with on the table.

An hour or so later, as the pyramid is getting fairly impressive, a mariachi band comes onto the patio and starts taking requests. Now the Beast only knows one mariachi song, but he likes it a lot, a whole lot. He starts paying the band to play "Jesusita en Chihuahua" over and over and over until the rest of the patrons start getting really annoyed. Fortunately, he is soon distracted when the "Men and Women of USC" calendar models enter the place and sit at two long picnic tables about halfway across the patio from us.

Apparently there is a calendar shoot taking place on the local beaches, and four female and four male models are taking their lunch break on our patio. The four girls are absolutely gorgeous, like Playboy centerfolds of the day, sitting in their bikinis and chatting with each other. The male models are a whole 'nother story. They sashay in wearing Spandex speedos and blow-dried hair, sleek and oiled, with every muscle in their abdomen perfectly sculpted. It doesn't take long to realize they are far more interested in each other than in their stunning female counterparts. The boys start passing around a bottle of baby oil and oiling up their pecs and delts in what is obviously a labor of love. The guys are caressing themselves and making the appropriate sighs of contentment. We just can't resist.

Now, we are all clothed in *tourista* splendor. The Beast is wearing a Hawaiian shirt in a gaudy shade of life-preserver orange and covered with sunflowers, along with a long-billed khaki tarpon fishing hat. R.O. is in a bright blue Hawaiian shirt covered with pink and white flowers, containing so much material to cover his nearly 300-pound frame that it looks like we are sitting across the table from the essence of Hawaii condensed into a shirt, topped off with an African pith helmet. I, for reasons I can no longer remember, am wearing the shirt and hat from a Turkish army officer's uniform.

We start to make some less than complimentary comments about the boys and their baby oil, possibly questioning their masculinity, who knows—it was a long time ago. Although they can't quite hear what we are saying they can sense a negative vibe and start giving us the stink-eye and making their own derogatory comments about our demeanor. Just when it looks like the situation might escalate, I tear open my Turkish shirt sending buttons flying across the flagstones, pick up a can of Tecate from the table and pour it on my chest, and as I rub the beer into my chest I look straight at them and say in a loud voice, "Facilitates the tanning process." They take a long look at me, a look at 300 pounds of R. O., and a look at the Beast who resembles a mad werewolf even when he is not hungover, and in one motion all jump to their feet and are out of the place so fast that speed lines hang in the air like a roadrunner cartoon. The four girls find this hysterical, laughing the entire time they gather up their purses and wave goodbye to us as they follow their fleeing counterparts out of the place.

* * *

During our time at Trac-Tel, a number of colorful and bizarre characters gravitate around our efforts. The most notable are two members of its genius roster. The first is Dr. H. He is a brilliant innovator in the detection of light and its amplification. His method of achieving this is a device using optical flats that he and Fred have cooked in some sort of nuclear oven that gives the flats lambda over 50, insuring there is no fracturing of the light being bounced back and forth.

Eccentric, lots of corduroy in his wardrobe, very hippie looking and mad as a March hare. Jeff and I are given the job of escorting the good doctor to and fro. It is entertaining but conversation with the Doc consists of his postulating the endless possibilities of night vision without cryogenics, up to and including replacing a human's eyes with a Trac-Tel manufactured device. I don't like the way he is looking at us as if he expects that we will be volunteers for this.

Because we have been experimenting with CCD cameras as a sensor head we have made contact with a small company in Barstow, CA. This company is called Continental RPV and produces a line of one-fifth scale Soviet aircraft, remote controlled. They are awesome to see in performance—they even have a Hind helicopter version but it takes two to fly. This company had gone after the drone market for the military in regard to training air defense units. Under the old system Lockheed produced a jet-powered drone that towed a banner behind and the guns would fire at the banner. The guns would miss occasionally and bring down the drone. This is an expensive loss and the drone was on a fixed path across the range and didn't do much else.

Continental have a more robust program, which includes shooting the drone down, and it still costs less than Lockheed. They have trained their operators to fly Soviet aerial tactics with jinking and swooping just like a real aircraft. Now the gunners are shooting at the actual drone not the banner.

The drones cost a mere fraction of what the Lockheed monster costs, they get better training and the costs are down. So far they have locked up training at Fort Irwin, the National Training Center, which all maneuver battalions must pass through every 24 months. They are doing a bang-up business. Additionally, the owners have invented an air-powered pneumatic launcher that throws the drone into flight at the end of the 15-foot launch rail. Very innovative, but resembles a cannon on two wheels.

We have taken some of the cameras and fitted them to a couple of their drones and are using these as an aerial, tactical surveillance platform. It works great: we can get out about three miles with it and see real time what the cameras see. This is years before the Predator and other drones will become standard equipment, and very cutting edge. We are trying to sell the idea to the military, so Jeff and I go to see the head of R&D for the Department of Defense. The general looks at our video and proposal sheet and points out that the signals can be jammed. True, and if you add the equipment necessary to prevent that, you go beyond the payload that the drone can lift. He suggests that we add a monofilament fiber optic to carry signals for both the cameras and for flight control. It cannot be jammed.

They have a system under review that does just that, but it is a solid rocket design with an antitank warhead as its major component. It comes in a six-pack that fits in a jeep trailer and can go out 10 kilometers and look for activity, then be directed to pinpoint accuracy. To demonstrate this accuracy they actually flew one up the tailpipe of a helicopter drone. We go back to Trac-Tel and they start working on the modification. Since ours is slower and limited in payload we decide to use Dr. H. and his night vision to have a technical edge. He is busy working from home and at the lab with the other techno geniuses.

It's a long holiday weekend and on Monday I go to pick up Dr. H. at his house. When I get there, no one answers the door and as I peek through the windows, every scrap of furniture is gone. The house is stripped bare, and the good doctor nowhere to be found. The neighbors see me giving the house the once-over and come out to tell me that the government came and picked up Dr. H. and all his belongings on Saturday. No explanation, just that they were from the government and a caution to the neighbors to mind their own business.

Fred comes back and also listens to the neighbors, calls his contacts in the intelligence community and basically is told that Dr. H. no longer works for Trac-Tel. Our star is stripped away so we cannot add his exciting technology. The whole thing smacks of *Men in Black*. It is what it is, but we still see a vision for the drone and surveillance with oil companies, commercial ventures etc. We make a small dent in that market, but they make most of their income from the target business. Later all of our improvements and the zero launch system become commonplace in the DOD drone programs. We are glad to be out of the scientist babysitting program. I will always wonder where Dr. H. went. Is he still locked in some sub-basement in a government lab somewhere, being fed on Solyent Green and howling at the moon?

The other rather odd duck is a physicist who Fred hires to solve video and visual data in the RF range and get real-time video remotely. His name is Jake, and he is a younger version of Dr. H., complete with the madness. Bill hires a tenured physicist to interview Jake for the job. After about two hours the physicist comes out and tells Bill that he feels like he spent two hours at Einstein's knee. Jake is hired; however, he has some odd pastimes as well. He has invented a plant growth hormone or some additive that he produces at home and sells at flea markets and farmers' markets on the weekend. It comes in one-gallon jugs. It apparently does what it is supposed to, and he makes quite a bit of income from it. He brags that it costs him around 11 cents a gallon to produce since it comes from algae. He has vats of the algae in his garage and sales are so brisk he has talked the local stop-and-rob into converting one of their water vending machines to dispensing his one-gallon miracle grow for 10 dollars a jug.

The Beast is a germ magnet despite his robust appearance. If there is a cold to be caught, an infection either viral or bacterial, he'll get it. He comes into work, leaking everywhere, coughing, feverish, all the symptoms of the latest flu. Jake is in the office because he likes to come down for breaks from the lab and fondle the guns with a dreamy look on his face.

He takes one look at Jeff and runs out to his car and comes back with a jug of his miracle grow and suggests that Jeff drink a glass of it immediately. He is pouring a glass and explaining that the algae will reestablish the pH balance in Jeff's body, improve his electrolyte sequencing and a host of other science-based reasons for drinking the solution. Jeff isn't buying it and the good PhD holder informs him that he and his wife both drink the solution every day. This is not a good example to throw out because both of them, though brilliant, are almost as loopy as Dr. H. had been. The scientists and quark-obsessed crowd upstairs try to stay out of our radar pattern, and we continue to try and flog their designs.

Bill decides that we should go to Malaysia to attend a huge international defense show, so he arranges for us to take a marketing junket. The only drag is that we have to take the Brit with us. We are busy preparing for the expected rocket shot into the exotic East. During this time one of our least effective brokers shows up and wants to have dinner with us and the Brit over a proposed exclusive for Korea. He is an energetic, possessed, frantic man who reminds me of a weasel looking for holes in the chicken wire. We agree to meet him at an upscale restaurant in Irvine. This place is two stories of sheer elegance and Southern California glitter. The upper level looks down on the foyer and the edge of the dining room below.

We get a table right on the edge with a chest-high glass and metal barrier to the floor below. I've taken to calling him Ming the Merciless. He also has the annoying habit of singing Italian operas when he is in the heavy ponder mode. One time (so he claims) he hit a high note and part of his throat ejected itself; which is probably why he isn't premiering at the Met.

He starts off with this long explanation why he is the best thing since summer kimchi and produces his copy of a document supposedly giving him exclusivity. He hands it over to me. I scan it and it's pure drivel, but it has my forged signature on the bottom which I point out to him and demand an answer. He turns to the Brit and explains his anguish that I am not honoring his clumsy attempt. Jeff starts to slide his chair back when I grab the Korean by the tie and drag him around the table, crushing the Brit face first into his Caesar salad. Once I have him, emotion overcomes caution and I hoist him over the rail and hang onto his calves then ankles, as he dangles upside down over the diners below. He is shrieking and pleading, and I am screaming at him to admit it is all a lie, when I realize that I can't hold him. Jeff leans over and we collectively pull him upright and he runs off wailing. We get the heave-ho from the restaurant's management, and drive back to Trac-Tel. We all go upstairs because the Brit is having a come-apart. He goes in with Fred and relates the entire story. Fred listens calmly and soon the Brit comes out giving us both a frightened look and scurries back to his office. Fred asks us to come in. We come in. He calmly asks us to give our rendition, laughs a couple of times during the version. Then looks back at us and merely says, that was a good try, but the Brit is still going with us to Malaysia.

When we finally leave, we have to stop in Taiwan on the way. The Brit is full of the high-profile folks that he has to see and takes Miller with him while I shop for a nice funeral horse (a ceramic funeral horse is a traditional Chinese gift on the passing of a loved one) to give to the prince we are going to meet in Kuala

Lumpur as a gift. They get to the location where they are going to meet the head of the Taiwan police intelligence division. The Brit sees his moment of glory and slides to the front, trying to obscure Miller in his importance. The good general takes one look at Miller and runs past the Brit, wrapping his arms around the Beast and shouting for all to hear that Miller is his esteemed teacher from the US. It was the little guy at the IACP seminar that we had trained in Tucson a couple of years before, that wouldn't give up the shield. It's a small world and his nibs is left sulking on an asteroid. We get a first-class sendoff with police escort and the general coming along to lavish praise on Miller the entire way. The Brit just sits there and sulks.

We arrive in the capital of Malaysia and find the usual arrangement where the COO stays in the five-star and we are in a separate hotel away from the property. This is fine with us since my overwhelming urge is to ballistically debrief him, or skin him and use it for a resin bag. We are meeting with one of the many Tunkus or princes. This one is the brother of the last king. Malaysia has a system where the kings of the separate kingdoms or provinces get together every six years and elect one of their own as king. The prince is a great guy, immeasurably more since he tires of the Brit before the first meeting is over. Jeff and I on the other hand are invited into his inner circle and spend the wee hours and a weekend with him and his cohorts. They have their own private club, complete with a putting green in the bar. They putt not for money but where the loser has to drink a water glass full of Johnny Walker Blue Label.

By week's end we have filled our Rolodex with some of the best contacts in Asia. Tunku H. knows everyone worth knowing in the rice belt and beyond and we develop a fast bond. He introduces us to his best friend, who also happens to be a chum of the Sultan of Brunei. He also has the only licensed private detective agency in the country and has the Pond's Cold Cream import market locked up as well. There is no breaking the hold that the SAS has on security in the country, but we do well in Malaysia.

We also have a chance to play the Scottish Highlands golf course. This posh club had been built by the British during the colonial period. We had been asked to bring a set of golf clubs as a present for Tunku H. to give to the Sultan of Brunei as a gift. We are on the links with the Brit and one of the techy types we brought along. He is a nice kid but keeps a weather eye on us. He cranks one into the rough and then proceeds to bull his way into the foliage to retrieve it. He scruffs around and comes out, a huge double handful of balls in his grip and a huge smile. "Look at this, I can't believe all these balls were down there!" Meanwhile, the Malay with us is having a fit and rambling away in broken English about how that is forbidden. He keeps pointing at the sign, just to the left of where the techy had emerged, that clearly states in English, Malay, Arabic, French etc., "Do not attempt to retrieve your ball from the rough," with a picture of a king cobra in the strike position.

About that time the Brit hooks one into the brush from behind us and moves up to go retrieve it. Jeff and I silently move in front of the sign as he hops out of his cart. The techy ruins our chances of leaving something in the food chain and shouts out a warning at the last moment. He is believable because he is pasty white and hyperventilating.

This trip to wonderland comes to a crashing halt when the ever-fragile wolverine comes down with Dengue fever. It's a rush to get him back to the US in time to save his fun-killing self. It takes a few weeks, but soon he is cured and fit for duty again. The Brit has come to the conclusion that we are dangerous to his becoming the center of attention, except as the butt of some crude joke. The tech guy goes into the lab and regales the rest of the staff with our antics against the Brit. The lab rats loathe him as well. Fred has the best spot because we have accomplished what we set out to do, and he assures the Brit that he will keep a handle on us.

The lab boys slip up though, and load my computer with some app called MacMelt. I am sitting at my desk down in our office which we have taken to calling Cockroach Central. I have just finished laboriously typing out our after-action report, and hit the save and send button to send it up to Fred. The screen display starts to literally melt to the bottom before my eyes. I am incredulous since the computer and I are aliens on the same planet that hate each other. I lose it and reach down, pull a nine millimeter out of the bottom drawer and put a bullet through the screen and into the wall, destroying the evil demon-spawned technological nightmare from hell.

There is dead quiet in the building and then Fred's voice comes on the intercom, asking me to come upstairs but to leave my toy behind. They had installed cameras in our office whilst we had been on our Asian tour. The lab rats had crowded into his office and keyed up our office to show me finishing the report, and the subsequent actions. After the shot heard round the world, they have locked themselves in the lab with the bulletproof windows and door. The Brit fled the building as soon as I had started upstairs, by way of the fire escape. Fred commiserates with me over the incident; he assures me that he will try and get the lab rats to recover my document and suggests I take the rest of the day off and nods at Miller to go with me. I stop in the parking lot to pull out a switchblade, wave it at the window to the lab and flatten the tires on two of their vehicles.

Our last gasp at Trac-Tel starts with a positive note. Jeff and I go to Washington DC and we meet with upper-level executives from one of the largest security companies in America. This company provides everything from armored cash transfer vehicles to full-on security systems complete with rent-a-cops. Usually this type of job attracts the mid-level Feeb retiree or some other ASIS graduate climbing the corporate ladder. We are pleasantly surprised in John, who is a dynamo of energy and almost as sneaky as us. The other member is, to say the least, stunning. Martha is a sharp, late thirties, very beautiful woman with a sense of sensuality that overrides your caution button. She also has a wicked sense of humor and they collectively launch us into contacts that we will use in the future.

We return to Trac-Tel amid internal fighting between Fred and the insufferable little Pommy, lawsuits and recriminations and a downturn in production. Fred needs to trim some fat and at that point we are suet. He doesn't want to do it but we can read the writing on the wall. In the end we can't help ourselves—we decide to go out with a bang.

The Brit, besides being a sleaze job, has some fears that just cannot be avoided. We are told by Fred that he will have to cut our division; he gives us a couple of weeks and a severance and a hearty thanks for our efforts. The Brit lurks in the background with a contented smirk over the way things shake out. We had scared the shit out of him once before with a plastic spider. He doesn't have arachnophobia, his case is worse. He shrieked one time in Korea when they served him crab because it looked like a spider.

I have been preparing for this day for weeks. I have been paying the kids in my neighborhood to capture spiders. All types, doesn't matter if they are black widows or just common daddy-long-legs. I have a shoe box full of the critters in separate boxes by species and sometimes divided by sex when the females use the males as snacks. On our next to the last day Miller and I get in early and salt his entire office with the horde. In the desk, under the desk, in the executive wash room, next to the coffee maker, in the sugar. There are spiders everywhere. We are sitting with Fred getting our final check when the insufferable little twit comes in and goes in his office. There is silence for a minute or two, then the entire building reverberates with this high-pitched screaming. Everyone rushes to see what the trouble is. The entire office is crawling with the Arachnid tribes. They are skittering everywhere. The Brit is curled up on top of the filing cabinet in a fetal position whimpering in sheer terror. Fred gives us a final look of remorse and we head out the door, giving Christine a high five on the way.

CHAPTER FOUR

Remake of the Wild Bunch

Unemployment is a healthy state. It makes you become self-sufficient and creative. We have been released from Trac-Tel and are eking out an existence, occasionally picking up a consulting job or security-related function. One such gig is security for a B52s concert, held on the grounds of Lion Country Safari in Irvine. It is gone now but it was an open-air park with paddocks to contain wildlife from Africa. It was a popular venue where families could drive through and see the animals up close. R.O. shows up one day and invites us over to his hilltop sanctuary for a BBQ. The meat is delicious. I have eaten springbok when in Africa, and this tastes amazingly similar. R.O. is a bow hunter and I suspect that Lion Country is missing a few tenants.

We show up for the concert, and spend the evening keeping the enthusiastic from mobbing the stage. The highlight is watching three ninja-clad morons climb the outer fence and try to sneak in the venue from the rear. Their ninja map was off because they climb over into the rhinoceros enclosure instead of the feeding lane used by the staff to feed the animals. They make it all the way to the next fence and start to climb over. This one is to the lion enclosure. Two of them are perched on the fence ready to drop into the lion enclosure when three lionesses show up for night snacks. The two fall back into the rhino enclosure with their partner. The lions are coughing chow call to the rest of the pride, whilst the rhinos show up to see what the ruckus is. The three are running hither and yon with the rhinos in hot pursuit. The park security folks eventually rescue them. It's all very entertaining especially with the music as a backstop. Rodgers and Hammerstein couldn't have matched the musical score to the run-for-your-life scenes.

One of the contacts we had gleaned from our final IACP seminar in Houston, Texas leads us to a father-son operation out of San Antonio. The father, Alfonso, is the spitting image of a Spanish grandee and the son is a self-possessed, second-generation sprout. We had gotten a contract from them to train weapons for Cementos Mexicanos in Monterey, Mexico, which the Beast, Penguini and the Grand Master had conducted. A few weeks later Jeff gets a call asking about Evasive Driver Training for another Monterey concern. Of course he says, "Hell yes we do driver training, it's one of our specialties."

Jeff and the Penguini rent a Chrysler LeBaron from one of the rental agencies to test out various exercises for a driver training class. They drive out on a dry lakebed in the Nevada desert and are practicing bootlegger and J-turns, when the car flips over on its side. Penguini is halfway through saying "Good turn" when he finds himself suspended from his seatbelt looking straight down at Miller. "I guess I spoke too soon," he mutters.

They manage to push the car back over onto its wheels. There are a bunch of scrapes down the driver's side and the outside mirror is torn off, but the biggest issue is that the front left tire is shredded and torn off the rim. They go into the trunk for the little donut spare and find that the combination jack handle/lug wrench is missing. Now they are several miles out in the middle of the lakebed, stranded. The only movement they can see is a radio-controlled model airplane flying around about a mile or so northwest of them, so it is time to walk. They hike over and luckily the guy is fully equipped for off-road, so with his help they get the Chrysler back on its feet. They return the car with the concocted story of how the tire had blown causing the car to roll and the lack of a lug wrench had stranded them in the desert without water. Full of indignation, they claim that they could have been killed. The agency doesn't charge them for the damage, thank God for LDW, so it's a clean wash. They decide we can do the training, but we need a real driving professional to spice up the mix.

Jeff and I find the answer to our dilemma in the form of a bartender at our favorite watering hole. Gordo is a sprint car racer in his off time, so we co-opt him into a training program. He is a driving fanatic and on top of that he is a natural teacher, with the modulated radio personality voice, and an actual vocabulary. This will eventually become a lifetime association with its own twists.

We do a couple of driving courses in Mexico, then Alfonso brings in the big fish. He has a customer who is one of the richest men in Mexico. He has a personal security outfit but wants to increase it by 50 people. He wants them fully trained and armed. Some will go to his facilities, some to family members, household, etc. We sign the contract and begin to assemble both a team and the format for the training. There are three pillars of close protection. They are shooting, driving, and tactics. Each of those have skill sets and procedures that must be taught.

We decide to start with interviewing, drug testing, and polygraphing to get 100 candidates for a selection phase, which will be in a remote area of Mexico. We know there are many prospective "bodyguards" that just don't have what it takes, and we want to get rid of the weak sisters before the client wastes a lot of money training them. We interview and test about 2,800 candidates looking for 100 good prospects, we wind up with 73 who pass all the tests. It will have to do.

Many dropped out because of the drug testing, most due to the polygraph. We have a set of questions that all are asked to determine their veracity. When it comes to one set of brothers, former federal police officers, the polygraph operator comes in and tells us we have a problem. The three Alvarez brothers are distinctly different. The oldest is very Indian looking, almost Mayan; the middle brother is a huge good-looking kid who insists on standing at attention when he talks to you and is very polite. The youngest one looks and acts like Eli Wallach as Tuco in *The Good, the Bad and the Ugly*. We ask the operator what the problem is: did they fail the lie detector? He replies, "No, not exactly." So what is the problem? He explains. "They answered yes to questions like, did you ever beat a suspect? *Si*. Did you ever plant evidence? *Si*, and a host of others." We tell him it's not a problem—they are telling the truth, so they are on the road to redemption. The three are star students and later go on to important positions in the private sector. Good lads.

* * *

Alfonso has rented us a remote ranch outside of Bustamante in Nuevo Leon. It's perfect. There are 60,000 hectares of wilderness broken into smaller 10,000-acre lots. The ranch is perfect for us, it has a main house and outbuildings and we can be out of the public eye. The area is strikingly beautiful with a high desert environment. The terrain is essentially flat but there are huge mountain masses that rise up from the monotonous plain. These soar up over 2,000 meters and the environment on top is alpine forests and rolling meadows of sweet grass. They raise fighting bulls for the arenas in the high meadows. Isolated and in the clouds, with no natural enemies, they grow and develop their skills fighting each other for dominance, until the roundup for the bullrings. These are huge muscled animals and extremely dangerous if you are afoot.

The town of Bustamante is the commercial hub for the area, supplying the rural villages to the south as well. It has all we need. The paved road ends about a mile from our ranch and another dirt road runs another 400 meters until it is no longer passable except by trucks and four-by-fours, then gradually peters out into the whistling coyote country of goat and cattle tracks.

We have arranged to buy surplus military tents, cots, web gear, canteens, stove kits, generators, light sets, and humanitarian rations for 120 people for two weeks. We have ropes and equipment for mountaineering and rappelling, rope bridges etc. We have scheduled the equipment to arrive three days before the students. Our staff is split between logistics and trainers who also act as platoon leaders, augmented by a medical section. We have an odd lot of instructors. Most are ex-Special Forces or military, and some are there because of a martial arts background. Almost all speak Spanish. Two notables are Kato, who I had known in Berlin and had been with on Project 404 in Laos as a young buck sergeant. He has evolved into a noted name in combat medicine and emergency medicine. He is also *intense*—no, that would be too tame a description of Kato. He is in amazing shape, and has bleached his long hair nearly white and tied it in a ponytail, so that he looks like Doc Savage, the Man of Bronze, a comic-book character from the forties. He also has the bedside manner of Dr. Mengele. The other is a Czech refugee who had emigrated from Spain after he escaped Communism. He is an imposing six-and-a-half footer and is a professional body builder. He had been in the Czech Special Forces or *Spetznaz,* and is a whiz at ropes and field craft. His only drawback is that he requires twice the number of calories to keep his bulk going and insists on getting a whole body tan, so he often wanders around in a wife-beater tank top and really short gym shorts. I'm thankful that thongs haven't come into vogue.

The management/logistical staff is headed by Jeff, me, and Alfonso. We purchase all the equipment and Alfonso is tasked with getting it all there by the date we have specified. This is best handled by Alfonso because he understands the country and the *mordita,* local slang for payoffs, to get the goods there. He has gotten clearances to pass through the military zones and has greased all the right palms. Gordo the driving guru will be driving the lead truck and also acts as the supply side organizer with two guys he brought along.

We set off to Bustamante eight strong, with another four following in trucks loaded with equipment, to arrive a couple of days later. The trip down is uneventful for the advance party. We arrive in the town and then take four-by-fours out to the ranch. The ranch caretaker, who will be with us for the duration, is a heavy-set Mexican in his late forties with severe acne scars on his face. His name is Rudy, and he makes himself indispensable at getting our little summer camp set up.

The first day and a half is spent checking out the ranch and selecting training sites for the various venues, and a clear field to set up the tents and cantonment area. The number of discrepancies between what was advertised and reality isn't that great, but noticeable. The original property had been owned by a very wealthy individual who finished up his life against the wall in Bustamante in front of a firing squad during the upheavals in the late fifties. The ranch had been broken up into smaller sections and sold off to the winners of that conflict, one of whom rented it to us.

It was advertised as a six-bedroom villa with an additional four bedrooms in an adjacent *casita*. Full kitchen, fully furnished in rustic furniture, in photos the place looks like the *High Chaparral* movie set, complete with satellite dish and swimming pool. In reality the satellite dish dates from the sixties and looks like a SETI array. The leads and connecter box have fried in the heat over the years and it is totally non-functional. There is a pool and it is full of algae-laden water so thick you can't see what's in it. We do notice that there is something that breaks the surface occasionally when you get near the edge. We surmise that it is a either a large iguana, or a caiman, or multiples of both, so a fresh dip in the pool is definitely out of the question. The original owner had imported animals from Africa which we are soon to witness, so it could be a Nile croc as far as we know. The bloody pool is big enough to be a decent 'gator hole. We use the shed and barn for the student mess and storage and we move into the main house. It's not the lap of luxury but I have used worse in the past. We are comfortable.

Penguini, Jeff, and I begin scouting locations to set up training. Needing to cover a lot of ground in a short time we opt for equine assets. Rudy has acquired a small string of horses, so we are using them to explore the back country. We are basically setting up a mini Ranger course. We have planned lots of forced marches, rappelling, and rope bridges to assess skills and overcoming the normal fear factors. All designed to instill team work and identify who the leaders will be, and to weed out the weak, the fearful, and the lazy.

Penguini and I are on our trusty steeds one day in an arroyo, down from a set of cliffs that we want to use for the rappelling and climbing. The horses start to get skittish so we decide to seek high ground. Sometimes the bulls get to the valley floor so we have been warned to keep a sharp eye. We are on a ledge about 20 feet up from the arroyo bottom. Suddenly three big animals break out of the brush below us and move around the curve out of sight. They aren't cattle. They are bloody great elands, a bull and two cows. They flash by and Penguini turns to me and says, "What's wrong with this picture?" Simple enough: this is Mexico and those animals belong on veldt in Africa. It turns out the guy imported them as part of his hunting preserve. Over the years they have increased in number to a sizable collection of bulls and their harems along with calves. So much so there is an official hunting season, complete with licenses. I am sure that the locals knock off a few as well. That's about 800 pounds of dressed meat on each one. This is also the day that I find out my horse is only a horse part of the time. The rest of the time he is a mule, and not just any mule, but an ultra-stubborn, stone stupid mule.

We work our way up to the cliff faces where we are setting up the rappelling station. Jeff, Penguini, and I are free climbing the rock face, and we discover it's trash rock, all decaying granite, that comes off in big slabs if you pull the wrong crack. Jeff does just that, and a slab the size of a pool table and about a foot thick breaks off and tumbles to the arroyo floor. We make our way back, hyperventilating, to safety; scratch that site but find some nice sandstone cliffs about a mile away. We also find the most ideal place for the rope bridges: an area consisting of several steep-sided gullies and a humongous canyon that is sheer sided with a gap of about two hundred meters to put a single rope across for the litmus test of balls and effort. It's perfect. There is about a hundred-foot drop when we get the rope set up, and the entire floor below is filled with prickly pear cacti and Spanish daggers. Fall off, you die, make it to the other side or else. This is a perfect final test. Tomorrow is the big day for Gordo and Alfonso to arrive with all our equipment. So we breeze back for supper and sleep.

Sleep is impossible because Rudy is the world's champion snorer. This is not just snoring; this is the guttural, hacking, throwing phlegm snore that resembles a cross between a lion's roar and a Sasquatch with inflamed testicles. We wake him and move him several times but it still does no good. We decide to make him sleep in the barn. This also proves to be less than perfect, because this is where our larder is; and Rudy is also a late-night snacker, to the degree that a whole chicken is light eating. We finally have him move into town and only come out on call. He looks at us sadly as he leaves, like a basset hound that farts inappropriately.

The students show up mid-afternoon on the third day, and we still don't have the equipment. They pile out of the chartered busses, many wearing suits and ties and reminding us of reception at boot camp. They gather their luggage which consists of either cheap suitcases or quality luggage, depending on their social station. As a group we walk them up to the main buildings. The barn is empty, so we decide to bed them down that night in the barn and shed. It's packed but livable. To feed them we send the talented Rudy down and he procures chickens and rice, and we cobble together a cooking station and hand out the chickens. The country boys soon have the meat processed and cooking, the city slickers don't fare so well. We see chicken that night that resembles a burnt shoe on a wire and some that isn't recognizable at all. We tell them this is the beginning of their training, give them a speech about working together and leave them in the care of the appointed platoon leaders.

Alfonso and Gordo are late, arriving near midnight. At the turnoff to the ranch, which is up a driveway some quarter of a mile, is where the dirt road ends. We agree to meet them there. You can't see the lights of the house from there so we are carrying a couple of Tiki Torches we found at the house and flashlights. There is a goat herders' hut with compacted floor of goatshit and mud with little sleeping platforms for the shepherds and their girlfriends in the flock. It's primitive, literally sticks driven into the ground to form the walls and a thatch roof. We decide to play a joke on Alfonso and Gordo when they get here by telling them this is the main house, not the advertised bullshit. When they arrive we tell all three to grab their personal gear and we will bed down for the night. We lead Gordo inside and R.O. points to a shelf and tells him to make himself at home. Gordo has arrived with full toiletries including a hairdryer, gallons of milk, snacks, small color TV, microwave, and his kit. He stands in the middle of the hut and looks pathetic. Alfonso is incensed and slightly less taken aback. We let it soak in as Gordo tries to clear an area to put his shit down, scaring up a couple of rats in the process. The rats scurry out a hole in the wall, looking back at us with promises of their return. Finally we relent: we tell them about the main house, load them back up and move the trucks up there. Gordo is so relieved that it's an actual house with fridge and power, that he overlooks the pool and satellite dish until the next day. We have at least patched the water heater, so we have plenty of hot water for showers.

We start the next day with the erection of the camp. We have four GP medium tents that will hold 25 men each and smaller tents that hold four. The commercial tents are relatively easy, the military tents take coordinated effort. We start by having them clear the area of brush and large rocks. Then we lay out the tents. We have people gathering small stones and gravel which we will lay as floor in the tents and sidewalks. This is as old as the Roman legions, everyone has a task and by mid-afternoon our camp is erect and functional. We begin to issue equipment. They all get two sets of fatigues, socks, web gear, canteens, and gloves as well as a rucksack and a numbered hat. Each of the three platoons has a different colored hat, orange, yellow or green. This makes it easy to tell where a student belongs, so they don't get mixed up at the training venues. Additionally, each hat has a number, 1 through 25, to identify the specific student. We know we're not going to have time to get to know them all, and 23 of them will be gone before the end of our stay anyway.

As we are doing this, we find out why Gordo and Alfonso were late. Their trip would have made a good Bogart film. They had taken the back roads after they cleared the border, Alfonso had done a route recon and paid off the military checkpoints to insure they could get through. As they were traversing one mountain pass, they had come across a checkpoint. This one had not given a rat's ass about Alfonso's pass signed by the area commander. After they opened the back and saw the military equipment they hauled them both out of the truck and forced them spread eagle on the dirt road with rifles pressed to the back of their heads. This was a time when the Zapatistas and other groups were using the area as their staging areas. The military thinks they are hauling equipment for the rebels. Alfonso finally convinces the young *tenente* to call headquarters. He does, gets his ass chewed, and returns to tell Alfonso to get his gringo and get moving. As it was, they spent several nerve-wrenching hours being detained which is why they were late.

Gordo, who had been skittish at the thought of being in a foreign country since we hired him, had thought they were going to be shot. By the time he arrives at the ranch he is a frazzled mess. Our goat shack had almost crushed him. He is recovered now and in full mode of improving our living conditions and acting as the supply chief. Basically, improving our living conditions means figuring out how to get TV and internet connections. He is a constant source of entertainment. He has brought six gallons of milk, which are in the fridge, and chained together with a bicycle chain. The big Czech doesn't see this as a challenge. He pulls them out and drinks from one whilst cradling the rest as a counterbalance.

We have a brilliant radioman as one of the instructors, he rigs up an old box spring and some antenna wire and has managed to pick up both Mexican and US stations. It's really quite innovative, using the box spring as a reflector, able to use ground and airwave propagation, and a host of other gibberish. I ignore it since if Gordo can get *As the World Turns* and the game shows that he is addicted to and the news, he is calm. We have far too much entertainment in the students.

Our intent is to weed out the weak sisters and produce a disciplined and bonded team by putting them through hard physical and mental conditioning. We start by making sure their rucksacks are full of the equipment that we issued; with extra water they weigh in around 30 pounds each. We have PT in the morning, training all day followed by martial arts before supper. The meals are all Humanitarian Rations similar to MREs. But these are mostly lentils, lentil soup, lentil stew, lentils, and rice, etc. and one that has ham in it. Ghastly stuff. We take to calling them Meals Rejected by Everyone. They are filling but if you are used to a hot pepper diet, the shock of bland taste soon produces the runs. We have to adjust the diet so Rudy goes on a constant run to keep us in peppers and some form of meat. Luckily, we are in a ranching area.

The instructors are cooking our own so we are fine. We are rotating cooking duties so most of the time we eat well, since most of us are fair cooks. There are a lot of homemade chilies, and stews. There are a couple of us however, that can't boil water so they are off the cook duty except for cleaning and fetching.

We walk everywhere, well the students do anyway. Usually the instructor/platoon leader walks with them. Penguini and I are on horseback and we have a couple of four-by-fours to haul equipment and water to the training sites. Our days and nights are full. We do tactical training to get them used to working as a tactical unit. We have paintball guns for part of this so they get used to handling something that can actually hit something at distance, and of course the ouch factor to remind them to move, shoot, and communicate. Most of the mob does well at the practical exercises, some don't. These individuals will fail to make the grade and will be cut. We also have a culling procedure: if someone fails or becomes an issue they are taken in lots to the bus station and released. The faint of heart go first, along with the slackers, or bad attitude group.

The culling is about equal, but the big Czech's platoon is down to 16 out of 25, faster than any of the others. Since we had divided the people randomly, so each group was more or less equal, we are concerned because we think that he is enforcing the rules too strictly. That isn't the only reason he has been cutting his roster. Since each platoon leader draws and issues the rations from his supply, each man has allotted rations, therefore when he cuts a man from the roster he has his rations. He needs more calories to sustain that body, so he has solved it in the most Soviet of ways. Moreover, he is only eating the ham rations, leaving the lentil crap for the troops. We put a stop to that, but the *Spetznaz* is constantly on the prowl for protein.

I buy two large summer sausages, eggs, and potatoes for breakfasts at the local *carneceria* in town. While I am staring at them through the glass of the meat locker, the light fades and I realize that something is standing behind me. I hear him say "Those are beeootiful" and then he is gone. I stash them in the fridge at the house and go to bed dreaming of a hearty breakfast of *huevos whateveros*. Next morning

I awake to the most wonderful smell of breakfast being cooked. I head out to the kitchen and there is the Czech, with the largest frying pan we have, filled to the brim with frying meat, peppers, eggs, cheese, and potatoes. I walk over and recognize my groceries, but figure he is making breakfast for all: simple mistake. I comment that his concoction won't feed everybody and that we should start a second batch. He looks at me blankly, and I suddenly realize there aren't any more groceries. It's all in the pan. He says he is cooking this up for himself, in all innocence, because up to that time only the milk was off limits. That is the last straw—his house privileges are yanked and we take away all the ham rations he has left.

* * *

The days are full of activity, and we are finally where we can start giving them confidence drills. Working from high places is excellent for just that. We have set up rope bridges over a series of minor gaps in the twisted terrain. These are along a preplanned route. They go to each station and learn how to tie knots, then anchor points then they construct a two-rope bridge then a three-rope bridge, and eventually will end up at the single strand that we have spanned the large canyon with. We stick to the first two bridges and one that we constructed that is a three-rope with slats for footing, and use it as part of the obstacle course that we have marked out. This goes well and we are soon timing them through the course. Our *piece de resistance* is the long one-rope strand that must be negotiated in the ranger crawl using one's arms, feet and legs to traverse it.

Jeff is running the obstacle with Gordo. The students come up and are given a chance to negotiate a separate 25-meter strand that's about six feet off the ground. Once assured they get the concept, we move them over hook a separate safety/retrieval line to the back of the harness, and snap link them into the main rope. We position them on the rope and tell them to start moving down the line. We also have incentives for fastest time, fastest platoon etc.

It is a chokepoint and soon we have more people waiting. We decide to take the ones that are frightened, or too out of shape, or just plain unable to coordinate the movement, on the six-footer and make a separate platoon for the exercise. We put in extra work on the six-footer and that eases the bottleneck. This gives us the chance to get the slow movers up to speed, and soon we have groups on both sides of the canyon shouting encouragement and derision at whomever is on the line. Our intent is to put a polish on them due to the danger factor. The canyon walls drop straight down a hundred feet below the rope. Most are hesitant getting up on the rope, but once they get out there, the rising shouts of advice and derision help them get across.

Jeff is at the anchor point and we are done with the ones that are going to get it the first time. We have about 12 Bolos, but we are determined to get them all across.

The Beast has come up with a method to encourage them if they freeze up on the line. He has brought one of the paintball rifles and if anyone freezes up he zaps one in their ass as an incentive. We get down to the last one, who refuses to give up and wants to do the rope. He doesn't want to get cut. He gets out on the line about halfway and freezes, Jeff zaps him with a paintball and he yelps then falls off the rope, held only now by his harness and the snap link. He hangs there limply, Jeff shoots him again, and again he just quivers when he is hit. Jeff switches to shooting him in the groin; still no results. Finally in frustration he pulls out his buck knife and screams at Baby Huey hanging on the rope, that he is going to cut the rope since we tried to reel him in but his clothing is in the snap link.

Gordo is wide-eyed and trying to talk to Jeff, who starts to saw on the rope with the knife, then alternates with hopping on the forward end of the rope like he is trying to break it free at the cut. The Huey scrambles back on top of the rope and starts down it, slow, deliberate but doing the process until he gets to the other side, where he is greeted by the mob as if he was a soccer star.

I don't get to see the whole event because I have ridden down with a party of students to the water point. That's when the horse for some reason decides to become a mule. He digs his hooves in and absolutely won't move. I lash him with the reins, I kick him in the ribs with my heels, and finally get off him and try and lead him, or drag him down the hill. No deal, he stands there and stares at me in mute stubbornness and refuses to move. I am getting more and more frustrated, especially since the students, as well as the Beast and Gordo have now stopped everything just to watch me. I search around and find a stout piece of mesquite and smack him on the rear, breaking the stick. He tries to kick me. Now I am enraged. I find another, larger stick about eighteen inches long and three to four inches in diameter. I walk around front and he gives me the bared teeth threat so I haul off and club him right between the eyes. The stick is completely rotten and explodes into a huge cloud of tiny splinters, which doesn't help my attitude at all. But even though the horse isn't really hurt, the shock seems to have had the desired effect and he is back to being a horse again. I turn around and there is the water party staring at me in horror, but further up the hill the Beast and Gordo are laughing so hard they are literally rolling on the ground. Well fuck them. I head back up the hill where Jeff pulls it together and gets back to finishing up the exercise.

When I get to the top the word has spread that I had been readjusting the headspace on my mount, so the mob avoids me like the plague, afraid that I haven't got it all out of my system yet. We decide to break everything down so the next hour is spent dismantling our little bridge kingdom and getting the equipment and trash moved down the hill and loaded on the truck. When I get to the bottom of the hill with my now very obedient mount, I find we have another problem.

The big Czech has been running around all day dressed in just bicycle pants and he is collapsed next to the truck in the shade with the remnants of his platoon gathered around eagerly awaiting his demise. He is lobster red on all areas of his body except the bicycle shorts area. Kato the medic is informed and we use the truck to load his body on board, soaking him with water and trying to get him to swallow. Gordo hops in back and off they go to the main house. We have one truck and a four-by-four left so we load all the equipment on board. It is a little over four miles back to the main house and tent area. We form the students into platoons and start back. When the Czech was with us he would lead the pack, walking and looking at his pecs.

The Penguini and I are near the front and the students are bitching about the horses kicking up dust, the fact that they're walking while the *pinche gringos* are riding. The sort of bitching you expect from the troops on a hot day. We are about a mile out when we both dismount and tell them we will double-time the last mile. We get them moving and we are on foot leading the horses. It is a glorious run, one that Penguini and I will cherish forever. We aren't exactly gazelles, being more akin to warthogs. When we stagger across the gate line, we are still in the lead and no more bitching about the *gringos viejo*. When we get there we find out that *Spetznaz* has full-blown sun poisoning and has to be kept in the dark for a few days. R.O. has him properly smeared with grease and creams, and is hydrating him via IV. We discuss in a stage whisper leaving him down here until he heals up, with Rudy. He squiggles around making moaning noises, which we take to be objections.

* * *

Our last day we have a surprise for the troops. There is a huge cattle sink at one end of our area near the house. For the uneducated, these are huge depressions that are part of the natural drainage which are improved and then surrounded with gravel berms. In the rainy season they fill up with water and stay that way through the dry season as the cows drink it off and evaporation depletes it further. There are dozens within five miles. The cows constantly use them, stirring up the mud, leaving their own waste, and grinding it all into a soup with a water layer on top.

We march the whole crowd down to the cow sump and tell them to strip down to their skivvies. Then by platoons we march them into the pond. The water is mid-thigh to waist deep in the center with about three feet of that being a viscous liquid soup of urine, cow dung, and water. The orders are to start fighting and try to throw the other platoons out. The last man standing will graduate that moment. It's a bloody donnybrook, there is no mercy; there is freedom and graduation, or defeat and shit details for the next two days. The competition is steep, and we glean the mob down to 20 or so winners. We march them back in the soup: this contest will decide the lucky one that graduates today. I am standing in the edge of the soup with the official start whistle in my hands. I am giving them the hurrah speech and tell them when I blow the whistle they are to start. I hear

Jeff and Penguini yell and get my attention. I look up and every instructor has a whistle. They blow simultaneously and the mob moves in, jamming me down into the mud. I am manhandled up gasping and passed ass over tea kettle to the shore and ejected. Some wit has taken a picture of it. I look like a mud baby, covered in cow shit and mud. The only dry thing I have is the stogie I had been smoking, and it is still lit.

The melee whittles down to two students, neither of which can get the upper hand on the other. After letting them wrestle for about 15 minutes Jeff decides they are both winners. We form the entire group up and introduce the two winners who have insured their trip to San Bernardino, and then tell them that every Sunday we will do this and select another graduate or two.

We can see them calculating: 50 men to San Bernardino, one or two a week, another 25-plus weeks of this. The students have no idea we are leaving in two days, their commitment was open ended. That afternoon our final seven dropouts present themselves at the house, bringing our total down to the 50 we need to take to California for the training.

We spend the next two days breaking everything down and having a graduation for the selected candidates for phase two and employment. Miller has made up some very trick certificates with gold seals and ribbons. These will be cherished by the ones that made it, and they are still referred to as the Bustamantes. Some will go on to be major figures in security in Mexico. The next phase will be in the US, in San Bernardino, where we will teach them the three pillars. The ones that didn't make it go home knowing they have been challenged and participated in something special. We get to go home and get ready for the real challenging aspects of training up such a group. We leave at night which sets the proper end to the phase. The sun is setting and the land starts to cool, with a slight breeze as we pull out of history and head home.

CHAPTER FIVE

Assassins on the Hotel Roof

In our plan, we would test, harden, and prepare a set number of bodyguard candidates for hire by the client as a permanent force. With the Bustamante stage completed the 50 survivors are officially on the client's payroll and the client is picking up all other charges as well. The students really want these jobs as they pay as well as being an engineer in Mexico. They are highly motivated!

Penguini accompanies the students on two busses across Mexico on Route 450, Route 10 and Route 2 to enter the US at the Tijuana/San Ysidro crossing and then up to San Bernardino. This gives the rest of us a week or so to get to San Bernardino and prepare. We spend that time finalizing liaison with the county and the city law enforcement and our venues for equipment. We have divided the tasks again between training, and administration and logistics. The first phase will be driving. Gordo oversees this and has a full complement of gear heads to help him out. We are primary instructors for tactics and weapons training and we will assist the driver training whenever we can.

Penguini has gotten the job of ramrodding the effort, a perfect fit for a former first shirt. He has been given a couple of guys that Miller has dredged up. I accuse him of getting them off a park bench under a newspaper. One guy hero worships the Penguini, but unfortunately he is flawed. He shows up with his wife, who is a diminutive Native American from the Andean Pact. We have rented the Radisson in San Bernardino, where students and instructional staff will be lodged. Everyone has a nice room. Tim, the assistant, and his wife move in, cover the windows with aluminum foil and keep it dark all the time. I keep looking for a coffin filled with soil in the room when I happen to glance inside. To say the least the kid is weird, but he is willing if not completely competent.

We also have one of Gordo's gear heads as a runner. He is a nice kid, a little out of shape but capable and conscientious. He does have his off days when all his circuits seem to weld together, such as our last night in Bustamante, when we had gathered at the ranch house and had a proper farewell to the old hacienda, fueled by alcohol.

We were gathered in the main room with its pre-revolution décor from a forgotten era, going over the mishaps and total FUBARs of the training and individuals, sort of a *Canterbury Tales* on speed. As we wound down, Kato the medic suggested that we all have a final toast. He pulled out a bottle of Metaxas, the Mexican equivalent of brandy, that doubles as dry-cleaning fluid. Everyone had a snifter with a generous dose. Kato then proceeded to light all the snifters, in a misplaced attempt to add a sophisticated air to the night's festivities. Everyone waited for the flames to self-extinguish, falsely claiming that the flames have added a special taste to the Metaxas, sort of like setting fire to gasoline and claiming it's now diesel. All except Jim, Gordo's faithful minion. He put his hand flat on the top of his snifter to extinguish the flames. Physics took over and the red-hot glass welded to his palm. He was shrieking and running around the room with the glass burning into his hand. Kato intercepted him and deftly smacked the glass with the fireplace poker, breaking it. Jim now has a round circle burnt into his palm. We keep calling him Kung Fu after the scene in the old TV show where Grasshopper lifts the cauldron with his forearms and the dragon is burned into his flesh. His scar which impresses the crap out of anyone that hears his version of the events, where it was an initiation rite.

One of the jewels of the staff is a former Special Forces, who had been one of the platoon leaders in Mexico. Tall, competent, and talented, he will be a genuine asset to the training. We will later use him on a kidnapping rescue in Mexico where he will again prove his mettle.

Jeff has also brought our old friend and compatriot Dr. Al along as our interpreter/personal valium injection. Al is a sardonic, laconic, and several other "ics," person that lends not only a wealth of talent, but does it with panache.

We have co-opted the sheriff's department since we need to use the city to do our training on surveillance detection and the scenarios. The hotel staff should get a special medal for their cooperation and service. They all help us create a unique training environment that will impart the necessary skill sets on our students and produce the type of individuals and team needed for the job.

We also employ two other SF types to help with the scenarios. One is from San Antonio, and married to a news reporter on the local Spanish-language station. He is a bit bombastic but otherwise fine when he is sober, but a genuine nightmare when he isn't. He will later cost us dearly when he gets out of control.

Rounding out the cast of thousands we have four people who will be our role players in the form of the family that they will protect during the practical exercises. This includes a native Spanish-speaking psychiatrist, who is there also to observe the students and give us evaluations.

We are using the Sheriffs training center and track for the beginning phases of the training. It is too small, so we move the training to the recently closed Norton Air Force Base and are using one of the taxiways as a skid pad with 11 lanes going simultaneously. Gordy has the station operating efficiently as we go through the basics of driving. Slalom, reverse slalom, braking, J-turns, bootlegger turns, and general handling during defensive driving. Most catch on quickly and are soon progressing nicely. We do have some that have never driven before, so we are running a remedial station to bring them up to speed. They soon excel at it. We have two that claimed they have driving experience, and maybe they did, but it must have been at a demolition derby. They wreck two of the rentals and we give them their termination orders. We send them down to the airport, with the Penguini. They decide they don't like the decision and get froggy at the airport.

The Penguini on his worst day is someone that you would avoid getting into a fight with. On this day he is tired, and his last chore is to drop off these two bozos then home for a shower and feed. They get out of the car and start to give him a hard time. He stoically watches them, pulls their bags out of the trunk and throws them at them, telling them that's the end of the argument. Their assumption that numbers would make the difference is quickly dispelled as he proceeds to take both of them apart in a matter of seconds, leaving them bruised and bloody on the sidewalk. Two deputies respond to the incident, who are familiar with who we are and what we are doing. They just grin at Penguini and ask him if he needs any help, or is he going to pick up his own litter. The two are hustled to the gate and put on the plane. Somehow the incident gets back to the hotel, so now all the students are *very* respectful of the Penguini when he asks them to do something.

We now split the class into two smaller components. Gordo selects 15 students he thinks have the skills necessary to make good drivers and starts taking them up to the San Bernardino EVOC center for advanced training every day. They will progress into driving in convoy and start to do more high-speed drills, J-turns with multiple vehicles. If done properly the three or four vehicles look like a choreographed dance. If done improperly it looks like a goat rodeo. They have a few goat rodeos but manage to get the cars and students where we feel that we can start doing ambush drills on a selected area that the sheriffs have cordoned off for us.

Meanwhile, the remainder of the group begins getting bussed to Riverside, on a bus we purchased just for this class, to begin learning to shoot. The client has purchased pistols in the US which will be sent down to Mexico at the end of training for his crew's use. Ten days of CQB (close quarter battle) and 10 days of Advanced Driving and the entire class is reassembled in teams of dedicated drivers, guards and team leaders.

We are using paintball guns to do ambushes whilst we are teaching "working a principal." We teach them formations and immediate action drills where they have to respond to a threat. We constantly drill them with the fact that if they have to pull their gun and defend, they have already made at least one fatal mistake. Constant practical exercises create the muscle memory so that they act as a coordinated unit. We are becoming proud of our work product. The training is identifying who the leaders will be and what teams work best with each other and in combinations which we will recommend to the client.

A prime example of the program is our star student during phase one: a man in his late thirties, who is one tough SOB. He had the fastest time across the rope bridge over hell, excelled in the hand-to-hand exercises, and is clearly a leader. He is a little rough around the edges and looks as tough as he is. This is a man that you would want at your back in a gunfight. He will eventually become a section leader of his own crew back in Mexico. A couple of years afterwards we will learn that he had been killed. He was with his principal at a bank when it was robbed. He stepped up and took on the robbers single-handedly while directing his crew to remove the principal. He died in the gunfight but so did the gang. A brave man, who did honor to his profession.

The second example is from the other end of the spectrum. We had three attorneys that went through the course. All were excellent at the tactics and would make excellent operatives working with the family members and children of the principal, and we recommended them for that slot. One of them, however, failed miserably when we got into the live fire with paintball drills. He was the lead driver in a vehicular ambush. He had left his window open which is a definite no-no. One of the aggressors shot him through the open window right behind the ear. He had a complete emotional breakdown. He suddenly realized that his job required putting himself in danger and risking his life. I took him to the airport. I was sorry for his loss, because he had been good at the training, but if you don't have the *cajones* for the job, you need to not do it. We respected him for admitting it and were sad that he left.

Our students are becoming proficient at the job. We are constantly drilling into them *advance work*; advance work is the prime directive. They need to think out of the box to anticipate danger and act as a coordinated team. Situational awareness is on all the time; if you don't detect the threat, you must know how to survive the kill zone. The drills are all designed to hone that awareness and reactions to the threat.

We are doing one such drill where we send the advance team to the American Legion in Highland where we have permission to use the facilities for the day. We have taken 10 of the students to act as aggressors. I have positioned them around the grounds; a couple are acting as if they are landscapers, near the front with their weapons hidden in the tool cart. The rest I have spread out, inside and out. The ones outside are actually hidden under the foliage that we have covered them with on the hillside where the convoy with the principal will pull up.

The advance team come and are on their toes. Suspecting that something is amiss, they halt the convoy before it gets there. I go down with the Penguini, and we ask them why they halted the exercise. We determine that they hadn't seen anything, they just felt something was amiss. Natural assumption on their part, and partial paranoia; we send them back and the exercise starts.

Penguini is standing in the middle of the parking lot as the convoy arrives, The crew properly exit the cars, the advance team goes into the Legion where the principal is supposed to have lunch. We let them go inside. After setting their perimeter and the principal giving the signal, they get ready to leave. They have been as jittery and on edge as a virgin in a lumberjack camp, the entire time expecting to be ambushed inside the restaurant. As they exit the restaurant and are in the process of putting the principal in the car we spring the ambush. The guys hidden on the hillside take on the vehicles and personnel outside, the two landscapers go for the principal and the people we had inside attack from the direction they had come from. They are totally surprised but their training kicks in. They manage to shoot their way out, into the vehicles and take off, with cars going every which way until they exit in a perfect formation. The parking lot is completely obscured in the dust. We are holding our breath expecting to hear a catastrophic collision. The Penguini has disappeared in the dust. As it clears and they move off we finally see him become visible standing where he had been when it started, unscathed and intact. They had missed him, but he had a couple of very near misses.

We bring them back, count how many have obvious shots on them, and count the hits on the aggressors as well. They had suffered only four casualties, one of whom was the guy who quit. The aggressors lost eight. The cars were all shot but the aggressors lost. We cover this in detail, showing where fire discipline or lack of it, tactics and reaction drills had paid off. We change crews and do the drill until everyone has gone through the exercise on this terrain.

This is the drill. Situations when they are afoot, when they are in vehicles, all sorts of situations where their training is exercised in the scenario. We even ambush them in the hotel during supper one night. The hotel is great and lets us use a banquet room for the shootout. Mind you we have rented most of their rooms, but they do have other guests and we try and keep it to a low roar.

After the ambush in the hotel we are going to begin their final exercise where they will protect our role players for 72 hours. We give them the night off and are going to start the next day. We all take a break and organize for the final. I am up in my room when the fire alarm goes off. Naturally, the Fire Department responds, all the guests are outside. No fire. We rightfully suspect one of our students pulled the alarm. After locating the alarm that had been pulled, we question all on that floor. One guy at first says no, then that he was tying his shoe, and that he might have slipped and pulled it. We get charged by the city for a false alarm. We tell the students that from this point forward any fire alarms pulled will result in us arbitrarily dismissing two people, innocent or guilty. The guy who pulled the alarm gets a proper beating from his mates. And that ends it. We have another class a few years later where the same thing happens. Apparently, Mexicans are fascinated with fire engines and all the hoopla.

The excitement diminishes and everyone goes back to their rooms. A day or two later I am sitting in my room facing across the plaza to City Hall. As I am sitting there a helicopter slowly rises up from below, not 50 feet from the building and facing my window. It looks like that scene from *Blue Thunder*. This one is a Hughes 500 with Sheriff markings and there is an armed deputy and pilot looking right at me and the other rooms on this side of the hotel. I look at them, they look at me. A brief flash of the fire engine incident flashes through my mind, but I shrug it off as the helicopter ascends, up and out of my view.

Not five minutes later my phone rings, I answer it and it's Penguini. He asks me if I noticed anything and I relate the helicopter incident. I ask him why, and he says there are police units pulling up out front of the hotel. I ask him, where are you? He says he is at the mall across the street. I ask him if he was already there or did he go to the mall after they showed up. I don't get an answer, but my phone starts buzzing with a waiting call. I don't have to put him on hold as his line goes dead.

It's not one call, it's three, one from Gordo, one from Alfonso, and one from Tim. Both Gordo and Tim tell the same story: the place is crawling with cops at the entrance and the Penguini's assistant adds that a SWAT team has just pulled up. Both are in the parking garage next door. I answer Alfonso's call. The first thing I ask him before he can get going is, where are you? He's also off property, but he relates the same scene from where he is at. I ask him when he can be back and get a vague answer. Obviously, he can see the hotel so he should be back in minutes, right? I ask him if he has talked to Jeff and he says that he will call him.

At this point I am hoping that some distraught woman in her forties walked into the lobby where her husband and his secretary were registering as a married couple and in her grief had pulled out a large caliber cannon, then proceeded to shoot her husband, the desk clerk, the bell boy and a dog, then locked herself in the ladies' john. Or possibly some elderly couple had fallen from the sixth floor through the glass roof over the restaurant. Please don't let it be something to do with us.

I am gathering up my things for the possibility this is something to do with us, when the phone rings. I try to ignore it but it keeps ringing way past when it should stop and roll over into the voice mail. I stare at it. It keeps ringing. There's a knock now on the door. Not the soft knock of Lucita the maid, but a "if you're in there open now" kind of knock. The phone keeps ringing. I finally opt for the phone it seems less threatening. It's not, it's Miller. He tells me to get down to the fifth floor right now and that a SWAT team has our students in custody. I ask him where he is and he replies he is on the fifth floor and to not be a moron.

The die is cast. I will have to do this. I am thinking dark thoughts about the Penguini; somehow, I suspect that he knew all of this before he went for his mall stroll. I throw the door open in mid-knock. Great, it is Dr. Al, who takes one look at me and says, "You've heard?"

I tell him what Jeff told me, as we start down the hall. Just before we get to the elevator, he says that he forgot something and heads back to his room. I am thinking over waiting for him at the elevator. I bet the thing he forgot was a rope escape out his window. The elevator arrives I get in. I punch the fifth-floor button and I sink into whatever is erupting on floor five.

The door opens and I am faced with an armed SWAT team member in full regalia. Behind him and going down the hall, are all the students on that floor, sitting on their hands, backs to the wall, legs outstretched. On the floor in front of them are paintball guns which another deputy is collecting in a bag. I spy Miller down the hall talking animatedly with the commander.

The deputy looks at me and asks if I am with this group. I start to stammer that I was looking for my wife and got off on the wrong floor. I never had a chance. Miller spies me and shouts to the deputy that I am one of "us" and to let me pass. At this point I want to be one of "us" like I want to be discovered practicing animal husbandry with the neighbor's dog.

The event begins to uncock itself as the SWAT clears the floor. As it does we get the students up, confine them to their rooms and go downstairs to have a stirring conversation with the sheriff, the chief of police, and assorted red-faced, upset people.

The situation was this, as we learned during the ass-chewing. The mayor was having an emergency meeting at City Hall in his office that faces our hotel with its balconied front. The week prior the mayor and members of the council had received death threats from a Mexican street gang. They had been in the meeting and discussing with law enforcement about how to handle that situation. They looked out the grand windows that grace the mayor's office and not 100 yards away are very Mexican-looking people on the roof of the hotel running hither and yon with what appear to be pistols, or at least bulky black weapons. They are shooting at each other when the entire assemblage in the mayor's office hits the floor, seeking cover. The sheriff and the police chief and everyone else assume that it's the cartel, showing up to exact revenge on the mayor. That's when the emergency response goes out and the rest is history. Alfonso comes in at the tail end of it, long enough to hear that we will get a bill from the city, the county, and possibly the Girl Scouts if anybody needs counseling. Alfonso is deflated with the prospect of having to explain to *El Jefe* in Mexico that the exercise is going to have some change orders and lopes off to remonstrate with all and sundry.

Jeff and I are leaving when I see the Penguini emerge from the faux foliage in the lobby. He slithers over and asks us innocently what's up. Jeff wastes air explaining it to him, more to have it verbalized so he can properly store it, than for Penguini's edification.

I don't even talk to Penguini. I know he was aware of it when it happened, probably saw the idiots, then figured the safest place to be was at the mall. I want to smear him with doe urine and leave him duct taped in the middle of mountain lion country with a tape recording of a wounded animal cry on it. Remarkably we are ready to start the exercise the next morning. We gather everyone together and tell them that the schedule is still on and that those that were involved will have to deal with their employer when they get back. Alfonso is pacing and scowling in the back of the room looking like a bedraggled Captain Blood on the poop-deck.

* * *

Our three-day exercise is built around the concept that the client is visiting San Bernardino with his wife and daughter. While here he will attend two business meetings and take his family to a restaurant, go to lunch and travel to specific locations. Our role players with the head doctor as the principal will do what normal families do and the students' job is to protect them and try to be inconspicuous as possible. The doc goes for a jog every morning in the park near the hotel and his security will have to go with him in close protection and blend in with the *Jefe*. The wife and daughter go shopping, go to lunch etc.

We have some surprises that key around awareness. Jeff has hired a Hollywood make-up crew and they have produced an amazing appliance, a latex mask of a 90-year-old woman custom tailored to the attractive 19-year-old actress playing the "wife" of the principal. The effect of the appliance is amazing. We had practiced on getting it on and off, with another actress who would be playing a nurse.

Harkening back to our time with Elizabeth Taylor and David's other clients, we know all too well what a mind-numbing experience spending hour after hour shopping with a woman can be. We tell the actress playing the principal's wife to spend at least two hours or more browsing clothes and cosmetics in the mall—we want to make sure her protection detail is good and bored. She *shops*, I mean she shops with the intensity of someone with an unlimited gift card. After more than two hours of following her around the team is almost asleep on their feet. They are semi-conscious at this point but still somewhat following procedures. After her shop-a-thon, the actress has instructions to use a specific ladies' room. Inside waiting for her is the other actress in full nurse's regalia with a wheelchair and the old lady appliances and costume. The principal's "wife" quickly changes clothes and the nurse applies the mask, wig and other appliances and in about four minutes pushes the wheelchair with the 90-year-old woman out of the ladies' room, right past the protection detail and to a waiting van just outside the mall.

The guards wait and wait, getting more and more nervous. Finally, they ask a woman approaching the ladies' room to check for their charge. She comes out and tells them the room is empty and now they are really flustered. One member of the team gets the courage to enter the ladies' room and sure enough it's empty. Panic sets in. Now the team is running all over the store and we finally take pity on them and gather them up. "Your principal left 20 minutes ago in a wheelchair," we tell them.

One of course says he knew all the time but had just enough doubt that he didn't want to accost the old woman. Sure! We point out to them that not every threat may be a violent attack. Lesson imparted is that you have to expect the unexpected.

The second surprise we spring on them that first night is at a well-known, hopping nightspot, where we will try and kidnap their principal in the back parking lot, having taken the precaution of having deputies isolate that area. We are paying the deputies and the city double-time to be there. They are happy. They get to make some extra cash and watch an entertaining exercise, and we are protected from some Bubba rolling up and deciding to be the Lone Ranger with real guns, or so we thought.

We have been frequenting this establishment and one of our instructors, the one married to the Spanish-language news reporter, has been chatting up the owner. What we don't know is that he has told her that he is with the CIA, or some other secret squirrel organization, and that we are conducting an exercise of national security importance. Yada, yada, yada, all of it was total bullshit. The drunker he gets the taller the tale. Finally, the owner is suspicious enough to have mentioned it to someone who calls the chief's office.

Our exercise also includes what we call the honey trap. This is in the form of a vivacious young lady who speaks fluent Spanish. She is waiting to go to the police academy and jumps at the opportunity. Her job is to get friendly with the guards and see if they will spill the beans on themselves, the client, anything they shouldn't be telling young ladies in the hope of getting in their knickers. She shows up for the night in the most alluring red dress that fits her like a second skin; she is, in a word, stunning. We get a couple of guys that spill a little, and one who tells her his history all the way back to when he was chasing donkeys as a kid. He tells her everything.

The scenario trundles along, with us monitoring the whole shebang outside. We take the principal out, we try and kidnap him, do a debrief and send the crew back to the hotel and replace them with a new crew. We keep the groups separated when they get to the hotel with Dr. Al monitoring them.

We are finishing up and are getting ready to break it all down when our deputies get a call that there is a disturbance at the bar. They crook an eye at us and we just shrug, hoping it isn't ours. We know that the last students just left so we are pretty sure we are in the clear.

The secret squirrel was supposed to be inside with another instructor, watching and noting actions inside. As we round the corner of the bar, there is a milling crowd of people in the center of which is the squirrel, surrounded by deputies. He is roaring drunk. He is telling the deputies mostly gibberish, but they aren't buying it. We go over and speak with them. I tell the idiot to go sit in the van until we sort it out. We get the situation calmed down, with the end result of us being eighty-sixed for three lifetimes. Jeff and I walk back to the car and tell James Bond that he is fired right there on the spot. We escort him back to the hotel, gather up his kit, drive him to the airport, drop him about a half-mile into the parking lot and tell him that he better be sober when he gets to the terminal. We leave him standing on the hard concrete under the halogen glow of the street lamps. He looks like someone abandoned at the end of the Trans-Siberian railroad. Dumped and waiting for the NKVD to come take him to the gulag.

Whatever we did is nothing compared to when he gets home to mama. The story got back to her and she has a come-apart. She calls us a few days after we finish, demanding to know why her honey has been fired. We keep it pretty mundane, saying that he has a drinking problem. The shrink was there, and he offers to talk to her. I don't know what he tells her, but she's looking for a used castration knife from a sheep operation when he gets done.

One of our own had fallen for our honey trap. He had been so intent that he was scaring off the other students so that in the end we only caught three that way. But it was enough to set the example of what we were trying to emphasize.

Our role player, the shrink, has been going to the park early in the morning accompanied by his personal security team, shadowed by a couple of our instructors. On the last day of the exercise we spring our surprise on them. They have been as nervous as a mail-order bride, every day expecting us to try something when they are away from the hotel. We had required them to mount a 24-hour guard on the principals at the hotel. They have this place sewn up and tight as a drum. Every time they slip up, we do a penetration or shoot one who is outside alone, instead of in a two-man team.

This morning starts out the same as every morning: the team and the principal show up downstairs in jogging clothes and form up, sending out an advance team as they get ready. We are using the side entrance this morning instead of the lobby. This accounts for the nervous sidelong glances I see as I come downstairs. Nothing different there, we have been accompanying them each day. The core team had been taking some obscure martial arts style course, paid for by the customer, who really wanted to see if he was getting his money's worth.

We have arranged with the sheriff to hire some of the deputies to basically attack and try and kidnap the principal. Their principal mission was to beat the crap out of the guard detail, to see how they would handle a physical assault, a real full-on defend-yourself assault. When we had asked to speak to the deputies and ask for volunteers, we had been in the watch room. We explained that it would be a controlled exercise and what the parameters would be, and we wanted the park to look about the same when the target was in the box.

We also told them we would give them 500 bucks apiece for what should be maybe two and a half hours of work. We have 10 grinning volunteers. One large specimen, very white, looks like a cube of muscle with huge arms. His equally large partner, Hispanic and looking like a larger version of Danny Aiello, the scar-faced actor that always plays the bad guy, looks at us and asks if what we were offering was 500 dollars in cash apiece to beat up Mexicans? I am thinking of a way to put it in a more professional context, when they both grin and high five each other. We learn later that they are part of the SLAM team that does raids and felony warrants. Nice lads. There is another one almost identical to the tank twins, who asks me if he can wear his assault gear and helmet. I tell him he can come naked with a barbwire club if he wants. With a kidnap team formed from our guys and students we should have about 25 people including observers in the park.

We have stationed ourselves overlooking the park where we can control the action (sort of). The deputies and our kidnap team have radios, and as soon as they are ready, we will start the group down and into the park. The victims can sense that something is going to happen, and the formation is nervous and tense. One of the radios squawks and it's from our observer on the far end. He needs some help because there are a group of people about to enter the park. It's a group of young men of color with a basketball. Penguini, sensing the need for haste and probably because he wants a better view, starts off in that direction. I start the victims off right behind him. They are moving fast because they are anticipating the recoil and are soon gaining on him. Penguini senses something is amiss and glances over his shoulder. His eyes squint, his shoulders hunch, and he picks up the pace. The security detail picks up theirs to match his. Penguini glances back, only this time at us. I give him the finger.

Never pass up the chance for comedy. I wait until he is abreast of the ambush and key the mike and say "Go." Miller looks at me for a moment as if to say good idea, or WTF, but only articulates "This could get bad."

The park has numerous homeless people that camp out in its shrubbery and wander around looking for the morning scratch and pee location. I notice that there are a few more than normally and some of the regulars are up early.

Suddenly the entire park to the front of the formation and from the right flank explodes in a full-on assault by homeless people, a guy with a dog, and two guys that emerge from the reeds surrounding the pond, in full assault gear: vests, arm and elbow hard pads, riot greaves, steel-shod shoes and helmets, all attacking the security detail. Some of the homeless people are fleeing in the direction that the Penguini was heading. They are impeding any serious escape route in that direction and the security team is dealing with a Celtic war charge. The detail collapses into a tight knot, which is getting beaten to a pulp and thrown to the ground and stomped.

The Penguini has his finest hour. I wish I could have filmed it. He is a superb martial artist and moves with a grace that is amazing for his size and shape. He turns on a dime, starts to avoid the herd of homeless heading for the far side and is confronted by the charge of the tank twins and their armored comrades. He spins and ducks under the twins' forward motion, does moves that would make Jackie Chan green with envy; there is a spinning side somersault, a leap, followed by what looks like ballet moves and he is free of the berserkers, he gets up the hill and is hopping from one foot to the other, bleeding off the adrenalin.

The detail has broken up into individuals and small knots of people fighting for their lives on the victims' side and pure adrenalin rush from our attack force. The only guy still fighting alone from the detail has taken off his belt with a huge cowboy belt buckle and is using it as a battle axe, swinging it and holding two attackers at bay. So far, we haven't witnessed a single flying heel kick, no knee breaks, nada. This is a brawl, there is no room for fancy martial arts, the only moves that even resemble martial arts are the Penguini's escape from the human wave.

When the front and flank collapsed the three-man detail with the principal did what they had been trained to do: they took the principal in the safest direction which was to the rear and back toward the hotel. Our kidnap detail is ready. Before the detail and the principal get to the appointed zone the two guys manhandling our good doctor are part dragging, part pushing him rapidly up the hill along the sidewalk. He is suspended in between them, when they try to negotiate a street post. Unfortunately, the good doc is in the middle as each chooses a side. They run him right into the lamp post, knocking him nearly unconscious. They grab his limp body and drag it in the direction they were going, he's slowly re-gaining full consciousness and struggling as they pass our second ambush and make it to the hotel lobby where they pile in with their bloodied charge.

The lobby is full from the late check-out crowd and the early brunch herd. They are confronted by a hyperventilating, gasping foursome, looking like survivors of the apocalypse, one of whom is bleeding copiously from the head. This starts a panic in the lobby.

At the same time police units show up at the park, because residents overlooking the park from the bordering apartment complex are calling in with reports of a riot. The SWAT team and the police show up in a slow flood until we have blue-light displays everywhere.

Our law enforcement guys had briefed the watch commanders for the night watch, and they had even paid the homeless, who were to flee the park so they didn't get injured. It wasn't much, but it was incentive enough to stay under their newspapers and castoffs until the fight started, then move to the far corner where the Penguini was trying to get a radio to threaten us with eternal damnation and physical harm. Hey, 20 bucks goes a long way in that environment.

The only problem was that there had been a shift change and the day watch hadn't gotten the word, so they had no idea what was going on. As soon as they realized that half or more of the combatants were their brothers, it went immediately from dire to near festive as our war party bled back into the assembly area complete with bruises and a few cuts and abrasions, still hopped up on adrenalin, giving each other high fives and chest bumps.

This is the final exercise, so we break everything down and get ready for the graduation which is held in one of the banquet rooms. Everyone is showered, bandaged, and assembled. We stand on the stage and each man comes up, gets his diploma, then goes down the line and each of us shake his hand and congratulate him. They are all so very proud. I would be too, they have completed an arduous course and are moving into a job that pays extremely well. We have a few that didn't make the cut, but they have been driven to the airport before we graduate the rest.

The diplomas are impressive and we are to learn a few years later that in the security industry of Mexico, to be a "Bustamante" as they called themselves, was considered a high honor and certification for jobs and promotions. Most of our students will go on to be security managers for large corporations and wealthy families.

Of course, there are always two sides of every coin, which means there is always danger when you teach these skills. Two of the three Alvarez brothers became major figures in security. But a few years later Jeff called me and told me to turn on the news. I turned it on and there was a short news clip of the arrest of a major figure in Los Zetas. He says watch when they bring the guy out. I am watching when they bring out the man. It is the youngest brother. We must have taught him well enough that the Cartel considered him valuable.

The saddest part, which resulted in this fine group being broken up, was that after three assassination attempts on the principal's life after he got the trained crews, which were all thwarted, the client went for a ride on horseback with his daughter on their estate and a sniper killed him from about 500 meters. They broke the team up and each went his own way. Even today they are known as the "Bustamantes."

As a closing note, we are all leaving the hotel to go home, Jeff and I are riding together when we pull out of our parking space on the second floor. As we turn we see Penguini in his Toyota Forerunner getting ready to straighten out for the exit. His minion Tim is talking to him through the window. We can hear part of the dialogue, where he is telling the Penguini how proud that he was to be wearing our uniform, how much he hopes that he can work with us again, etc. He is holding on to the partially open window. He never gets to properly finish his "ode to the Penguini" because Penguini rolls the window up on his fingers and starts down the exit ramp with Tim pleading with him to roll the window down. Penguini finally takes pity and releases him, then we speed out of the garage leaving San Bernardino in our wake.

CHAPTER SIX

Send Lawyers, Guns, and Money

'cause the shit has hit the fan.

—WARREN ZEVON

In the months following we have split to our own devices. In the interim I have been talking with a banker/bullshit artist about some consulting work. I am of the age when adventure normally takes precedence over common sense and mortality concerns. Jeff is the same way and so is the rest of our family of spiritual dwarves.

The Banker is a diminutive, furry bundle of nerves and big dreams. He is an almost carbon copy of our friend Jay, except where Jay is the embodiment of mayhem and nightmares, the banker's soul is a festering cesspool of deceit and manipulation. He has a plan that he must discuss with me, so he flies down from his lair in the Bay area, and I agree to meet him at the Orange County airport. I arrive and find him at the baggage claim, and we take the escalator to the parking garage. He is talking a mile a minute as we reach the car. We get in and I adjust the mirror. The minute I met him my hackles had come erect; something is not right. I had noticed two people that seemed a bit too interested in looking innocuous that had followed us out of the terminal. As I watch they are getting into a sedan that screams "government" as I back out of the space.

I had worked here at John Wayne Airport renting construction equipment during a dead period not long ago when they did the rebuild. There is a feed road with an entrance just before the normal exit from the garage. It leads across the airport and exits onto Bristol Street. I jerk the wheel and take it. As I look back there is the unmarked sedan on the exit ramp and the two guys are watching me disappear out of the airport.

I grab the opportunity to ask the Banker why he has an apparent police tail on him. He mumbles something about a tax problem but that his lawyers are handling it. I tell him tax problems don't come with a tail. He gives enough of an explanation that I take him to a safe house for the meeting. He has a project, and he wants my help to organize it. I almost shoot him when he blurts out what his scenario involves. In hindsight I should have.

During World War II the Japanese rolled through Asia, a deadly juggernaut and looting machine that was behind the Empire's expansion. During that time the army that took Southeast Asia, the Philippines and Indochina was led by a brutal general named Tomoyuki Yamashita, the "Tiger of Malaya."

His troops raped and murdered the conquered nations with abandon. They killed innocents in the hundreds of thousands. Many were Korean conscripts that followed the combat formations of Japanese regiments who were only a shade less brutal.

Every country that they conquered they looted the banks and everyplace else with valuables. They looted the Imperial Treasury in Nanking in the form of 75-kilogram bars of gold and silver bars worth millions of talents. Yamashita kept it with him and when he got to the Philippines, he decided to stash it in case the Rising Sun crowd lost the war. This was a massive effort, burying and hiding such huge amounts of gold. It took years to hide it and the caches were constructed by his engineers, complete with every imaginable booby trap, to include dead falls, explosives, poisonous snakes, gas, you name it.

There had developed a complete industry around the story of Yamashita's gold. Japanese participants came back after the war and opened mushroom-growing farms with sheds that hid the excavation of the caches, and professionals as well as amateurs roaming the hills and evading bandits in their search for this golden hoard. The sort of people that were around this gold were the kind of people that are the last people you see in this world if you make a mistake.

The Banker/schemer has been in contact with a local who claims to have some 1,000 kilos of gold in bars and pancakes. They have provided proof in the form of a doré bar cast with the proper Japanese seal. It is a known fact that the Japanese had cast some of their loot into doré pure bars and had stamped them. Some had surfaced before and been authenticated. The deal had been payment only on assay in an escrow to a bank in Manila. The gold arrives, they test it, and the gate opens on the other end for the money. It was a 5-kilogram bar which was worth about 75,000 dollars in those days. Now they had put together a consortium to buy the full 1,000 kilograms. That was over a ton to move just for basic logistics, plus how to get it to a depository where it is safe, while avoiding the Japanese secret police if the story is true and staying alive. Sure, why not? What could go wrong?

I am at least aware enough to ease into this and agree to help set up the meeting in Manila and make sure we have cover if someone starts shooting or robbing. We know a number of expats living in the country, so I have co-opted two of them to cover us while in Manila. As for anything after that, we will look at it and see if those legs are feasible then decide how much we want to bite off.

On one of my consulting gigs I had been helping a company buy some C-47s that the Philippines was retiring, then taking them and replacing the engines with turbo props, stretching the fuselage then putting in all new electronics and avionics. Shazam! You have a perfect mid-range hauler that can be adapted to many things. It cost about a million to do the refit but when they are done, you have a 30-million-dollar aircraft. I had been on the very fringe of the affair working for one of the agents. During this project I had met several high-ranking military officers in country and as soldier to soldier we became chums.

In the course of events I had been with General B. in his rambling mansion on the hillside. We were in the game room drinking when I, seeking a seat, sat down on what I thought was a pool table, covered with a green tarp. He came over a bit later to where I was sitting, motioned me off then swept the corner up. There they were, neatly stacked 75-kilogram bars with the Imperial stamp on them—Yamashita's gold. I didn't say a word to anyone but that was the only true fact that I witnessed. It was well known that the president and the military were constantly looking for the gold and had found some. Lots of some apparently.

I heard rumors later that a good portion had been smuggled out by pouring real dense cement around the gold then planing it into huge blocks the same size as the ones big hydro-electric generators are balanced on. They are periodically removed then sent back to Norway or some other Scandinavian engineering center. Only these never made it to the shaving, they are broken apart, the gold melted down and recast in hallmarked gold in separate depositories under separate accounts. Rumors had Adnan and the president drinking champagne on board his yacht, with all kinds of embellishments and sidebars.

I knew one thing and that could have been an illusion or a diversion. It could have been cement gilded with gold. I am skeptical but I have brought Jeff in from out in the cold and we are planning the trip over and contingencies. Jeff puts up with the Banker, but occasionally does the universal screw loose motion when he is talking. Midway to the takeoff date, the Banker approaches me with a vaguely related proposal to recover a stolen bank document. Jeff and I have developed a good relationship with a gent named Dick F., a former intelligence officer slash director slash knew everybody to cover our ass. So, we run it by him, and he does some research and establishes that such documents exist, how they are traded between banks and who can cash them. I am showing him copies of the document and he is advising us on how to authenticate it before we get our tally whackers in a vise. So far everything checks out.

The document is an International Bill of Trade issued by a Japanese bank against an agricultural cooperative group that controlled rice and sake production, as well as some heavy industries. The cooperative is rumored to be connected to the Yakusa, those nice gentlemen with tattoos and missing digits. The same group had been connected to the gold before, so we suspect where the Banker got the connection.

The value of the certificate was issued against the assets of the group and the denomination is in internal yen. It is not supposed to be outside the country, but it had somehow been brought to the States and someone, I suspect our furry little Banker friend, tried to use it to leverage land purchases in the path of the first

California high-speed rail link between Los Angeles and Vegas. That deal exploded somewhere, which I believe is why the Banker picked up his police escort at the airport.

The document then surfaced in Germany, where someone tried to cash it at a trust at the Bank of Bohemia (this time the Russian mob; a German count with a BKA (German criminal police) file as thick as a Gutenberg Bible; the son of the family that owns most of Garmisch-Partenkirchen; and a Ferrari dealer in Switzerland). What the crooks didn't know was that in order to cash the instrument you must be a bank, because it's a bank-to-bank document. You also need the rice paper proof that fits over the bill with matching water marks. They didn't have it. Now they want to sell the document back to the rightful owners, of course for a fee.

The owner of the document is willing to pay our rate for security and providing any and all needs, to be supplied through a Swiss bank. This is going to require a lot of people and assets to accomplish. It totally absorbs my time. When the time comes to go to Manila, I have to be in Germany to start organizing the exchange and tracking all the participants.

Jeff takes the helm on the Manila run. And I begin to assemble a team for the recovery and the proper conditions to make it happen. We are going to meet up after he and the Banker come back from the swamp. It should be simple as this is the eyeball-to-eyeball meeting where they are going to do another small exchange, only this time a number of larger ingots as proof and the arrangement for the exchange both at the bank and for possession of the gold. It will be a test run for the big kahuna. I still am full of worry about double crosses, and double crosses inside of enigmas wrapped in puzzles. Jeff can handle himself and bring the Banker back, assuming the Banker will listen to him. The Banker by now has raging gold fever and only sees the gold, not the brigands surrounding it. They are off to Manila flying business class and I start getting the European project off the ground.

Over the course of years, we have run into a lot of people on both sides of the law and those that hang on the fringes. One such contact is Gunther, a German resident who is an American citizen, married to a woman 25 years younger than himself. His father-in-law runs a butcher empire in Bavaria. He has contacts in Switzerland who have contacts with both the government and the banks. We are going to go see about luring him to Switzerland where we can do the exchange. This is what the bad guys suggested. Which I view with relief. Much better than some Bulgarian seaside resort or ships at sea. Hopefully we can control the terrain and have backup. My contact assures me this is the case. I have planned to pick up two other guys that I know can handle the job. Now we have to get the equipment as well, so we board a plane to Frankfurt.

I have a friend, the widow of a very good friend from the old days who died disconcertingly young. She had been recovering from his death in Australia, and had gotten a job working for a travel agency. Australians are geniuses at this since just going to the loo involves tickets and a bus ride.

She takes off to Mexico to review a resort for the agency, the same day that I leave for Europe. Everything is ticking along like clockwork. I rent a car in Frankfurt and cell phones to use, and Gunther and I drive to Switzerland and meet up with his contact there. The man is a major industrialist who has the right juice to get the document lodged in a bank for retrieval and the right connections to get the courts and the police to help us if it goes sideways.

When you meet a person and your hackles go up, try and listen to your instincts. Don't hang around people that dogs growl at. This guy is the epitome of deceit and just pure sick evil. He apparently "assists" the government and the banks to help get their money back when they are swindled. He has been doing this for several years. Swiss banks get robbed just like anywhere else, they just don't advertise it. They catch the perpetrators and strip them of their wealth then kick them back out on the street, or they get a plot in the old Stasi fruit orchards in the former East Germany.

I don't trust the guy farther than I can throw a buffalo, but he has the juice to get things done. The Banker in the meantime is selling me and all my playmates as a potential resource for his collection business. This is something I definitely do not want. I am in contact with the Banker by phone and I brief him on our progress.

At this point I get a disturbing phone call from Miller. Things have wandered off the path. The Banker has found a trust-fund baby, prep school grad, Ivy League hockey player, who was financing the entire trip on his credit cards. He maxed out Visa, Mastercard, American Express, and by the time they were done was almost maxed out on Diners Club. After they arrived in Manila our guys had helped them get the rudiments and marry up with the local contact. It was a weird triumvirate consisting of a young kid, obviously a little streetwise, his overweight, matronly sister, and an older man, an expat gringo, who claimed to have been one of the first Boeing test pilots for the 747.

The sister tells them that "grandfather" who controls the major stash of gold (which at different times fluctuates between one and 100 metric tons) doesn't want to come to Manila. They want to meet up-country at the northern tip of Luzon in the town of Tuguegerao, close to where the hoard is supposedly hidden and where they feel better about the exchange. This spells BANDITS in big letters. The sister, who allegedly owns one metric ton of the gold herself or something to that effect, insists the Banker front her bus fare to go to Tuguegerao to make the arrangements. The Beast has done his homework and discovered that in the Philippines a person can sell up to 8 ounces of gold a year no questions asked, so this woman needing bus fare runs up an enormous red flag. He is concerned that the ransom they can demand for the Banker is worth much more than the one bar they haven't been paid for yet.

In addition, the former Boeing test pilot is planning on flying the gold bars to Hong Kong in his de Havilland Moth which can only carry two bars at a time because of the weight! But Jeff learns that Hong Kong doesn't even accept overseas flights of single-engine aircraft. The whole plan is starting to stink like last week's fish. Jeff is telling the kid the deal is off, but the kid is adamant, and the Banker is so heavily in the grip of gold fever he is ready to head up-country. Jeff tells him no, not just no, but fuck no! He tells the kid that if the gold is in Tuguegerao then he is free to head up-country, bus fare paid by the Banker, and bring a sample back. "Bring back a gold bar, a gold coin, a gold tooth, anything at all and then we will all go up to Tuguegerao together."

Meanwhile, they wait around Manila, the Beast gaining weight from his prodigious pork adobo consumption and the hockey player keeping the ladies in the Australian-run red light district happy. The next day the kid goes on his own up-country to talk to the gold people and disappears. A few days later the sister calls in hysterics: her brother has been kidnapped and is being held for ransom. Jeff packs up the Banker and his trust-fund-baby financier and they head for the airport, leaving the country just ahead of the kid's head arriving in a sack at the hotel.

Jeff calls from Japan to tell me the Banker is a nightmare to manage, but at least he saved him from a Tuguegerao decapitation. They are staying for a few days in Japan because the Banker has meetings. I tell him that when he gets home that he will probably have a few days then he is to join me in Europe.

* * *

Meanwhile, I am looking for a good friend, Reek, who had retired from the military and moved to Germany with his new wife. He is a hulking, violent package, with an IQ in the Mensa range. He speaks German, Spanish, and probably Martian. I locate him in Baden something-something Stadt. He is running the property disposal yard for the military near one of the large bases. This is essentially the junkyard where vehicles that have been abandoned by their owners, or wrecks hauled in from the Autobahns, are towed. He has a lively business selling the parts and pieces and shipping the stripped hulks to steel mills after they compress them into flattened slabs.

My Australian friend calls me in midstride and informs me through tears that she is being deported from the US. It seems that she had returned from Mexico, and her visa had expired. She explains that she had mailed in the renewal a month before and had the stub that was her temporary visa until the new one arrived. She had paid the 79-dollar renewal fee with a money order and had the carbon of that with her as well. Leave it to a woman to travel with all that documentation, it's in their DNA. Guys can barely leave the house with our keys, our ID, and a condom. As she was explaining in indignation that she had followed all the proper timing and procedures, they informed her that the fee had doubled, and as a result her reapplication had been rejected. Now she would have to be turned away at the border, which meant going back to Mexico.

On top of that she would have to reapply in Australia since that was her nationality. They bullied her into accepting self-deportation, failing to mention that because she was past her departure date, she would be denied reentry for two years. I calm her down and call my good friend in Mexico who knows the owners of the Camino Real Resorts and she gets comped in at a five-star resort. She is still beside herself and distraught. I decide that the best thing to do is bring her to Europe and put her on the payroll as our secretary slash admin person that can do the chores that we haven't got time for, such as purchasing and transport.

I leave for where my friend is running his scrap empire and find him out in the yard. He sells parts off the cars for those talented enough to do their own wrenching. I walk out to the yard and see three people standing next to a wrecked Mercedes, and a pair of legs and shoes sticking out from underneath. I walk up and it's a major and his wife from the nearby base and Reek's new wife. As I approach his wife asks the major's wife if it's true that anal sex is illegal in the United States, followed by the statement that Reek has been teaching her English, and asks the major's wife if her English is understandable. The major's wife looks like someone goosed her and hides on the other side of her husband, who is intent on getting his auto part, and ignores the comment. I walk over and look down into the engine compartment and there is Reek looking back up at me. The only thing that I say is, the scene would be perfect if he was bald and fat since I last saw him. He isn't, he is still the hulking ogre I remember and looks even more evil in his grease camouflage.

He gets the major's part after he pays for it and the couple scramble out of the yard like someone had just asked them to participate in an orgy with orangutans. We retreat up to the office and he breaks out a bottle of Uppsala, the raw schnapps produced by the Teutons as a social opener and cleaning fluid. Over a couple of glasses, he lays out his empire and how it works. He is stripping the high-end cars of motors, transmissions, brakes, axles, seats, interiors, you name it, and loading these into 40-foot containers then shipping them to Poland, Russia, and even Nigeria. He is making good money and is hardly interested in my endeavor until he negotiates a better day rate from me. It's fine with me: I had been prepared to pay him double, because if it came down to deadly force issues I wanted him to back me up. We agree and he will be ready in the morning but wants to bring his bride along. I love his wife because she is funny and can help, but she doesn't need to come to Switzerland with us. He leaves her there to manage the yard with his two employees.

Reek and I spend the five-hour drive to Switzerland going over the plan for the retrieval and for the surveillance needed on the key players. We are lodged in luxury, which the Swiss are bankrolling. Thank God, because a beer half the size of an American draught costs 11 dollars at the hotel bar, bloody ridiculous.

The Australian is due to arrive on the morrow, so our first task is to pick her up at the airport, bring her back and give her duties. Simple, right? The next morning it is near blizzard conditions when we arrive at the airport. We wait outside the immigration gate and soon she appears dragging her matching luggage. It only consists of two small pieces, one on wheels and the other the size of a small picnic basket. She is dressed for Puerto Vallarta. Bright tropical colors, summer clothes, looking like an emergency beacon, they are so bright. The Swiss ignore her, as they are used to folks from the other side of the planet arriving unprepared for the season, usually on the lam with suitcases full of cash.

After the greeting exchange and the obvious relief she looks around and out the windows at the howling snow storm and states practically, "I'll need a new wardrobe." Simple, concise, and with full assurance that the wardrobe will be part of the job offer. We all go back to the hotel and get settled in. All the turmoil that had been dragging me askew was now here where I could deal with it hands on. Perfect. Let's get the job done.

Jeff arrives two days later with war stories of the great gold hunt to entertain us into the night. Gunther is a constant pain in the ass; the Swiss predation is even more so. He is an arrogant, scheming, backstabbing slab of *scheisse* as far as I am concerned. Gunter is one of those people for whom the solution to any problem involves criminal activity and is constantly trying to rush things in order to get at the money. This is not surprising since he grew up during and after World War II in the wreckage of postwar Germany, with baths once a month and selling cigarettes or purloined gasoline on the black market.

Our deal is that the bank pays for all expenses plus a job rate per man. On the backside we are charging a straight recovery fee of a few percentage points. At least we will get paid for our work. I am prepared for someone screwing us on the backside but have made it as ironclad as possible. The Swiss will rape the Japanese bank on a discount rate but that's their problem; we are to be paid in escrow upon receipt of the document before it is sent back to Japan. The Banker, back in the States, is only slightly more trustworthy merely because he hasn't robbed anyone yet in a deal this size. The Instrument is worth several hundred million dollars, so the backside is a comfortable bonus. The Banker already is incensed that I want a bonus for all the worker bees. I tell him that if he brings it up again, he won't need his share.

I have Jeff start the work on containment of our principal targets, those that have the document, and those that will pose a problem. Those that will pose an immediate problem are the Russians. As a group, Russians are direct people. If you are a problem, someone finds your remains years later or in the food chain of the nearest bog. I have Jeff notify the terrible Penguini and his older brother to be on standby. They will be handy if things go sour at the exchange and get nasty close up. Meanwhile I have Gunther arrange for some hardware. Reek and I start the process of contacting the principal players who are the German baron and his sneak twin the Ferrari dealer.

If we are going to follow these guys we need transportation, radios, satellite comms, and cash. The Swiss bank are generous to a point in my requests. They provide us with rented vehicles, cash for expenses, and after I discover that the phones that we were renting from the kiosk in the airport in Frankfurt were set up to back charge to Germany long distance, both receiving and transmitting, a complete satellite down station mounted in a Mercedes van with 12 slave units. They also provide us with a private jet should the parties hop on board a plane for somewhere, so that we can arrive before the commercial flight can land and pick up surveillance on them.

We make the first contact with them at the train station in St. Gallen, conveniently located in the northeast corner of Switzerland just a few miles from both Germany and Austria. This has all the earmarks of Russians. We are instructed to go to a phone bank inside the terminal, the fourth phone from the left will ring, and we will be given instructions on what to do next. Reek and I with Gunther in tow arrive early. We leave Gunther in a car with a monitor and we are wired up so he can hear all that transpires. We do a counter-surveillance sweep and locate at least three people that seem much too interested in looking inconspicuous. We go out a side entrance and circle back to the car, get different jackets and shoes then go back to wait for the call. Just like clockwork the phone rings. I pick it up and the voice on the other end gives me a detailed route I am to walk to get to a meeting at an outdoor café. They are doing this so they can determine if we have company. Reek goes with me paralleling my route with Gunther's help via radio, and we wind our way through their route until we arrive at the location.

When we get there, our contact is the German baron; the one with the long criminal record, mostly as a gigolo ripping off aging dowagers. He is a big man, florid, in his fifties with a paunch, decaying looks, and a fake tan. I move over to his table and ask if he has finished reading the paper laying on the table in front of him and he replies with the proper response. It's not all that hard to identify him as my contact. Everyone else in the place is wearing the ubiquitous work coveralls worn in the country, unshaven and oblivious to anything but their coffee and schnapps.

The baron thinks he is in a spy movie and makes all the mistakes of an amateur. He talks too long and too much which I am sure is vexing his Russian partners. We establish that we are with Lloyds, thanks to some paperwork provided by the Swiss, and assure him that we have come to retrieve the document and that we have the money to do the transaction.

The Russians had already tried to cash the instrument and know that they can't do it unless they are a legitimate AA bank, so they have opted for a straight ransom payment of two million Swiss francs. The baron gives me instructions for where the meet will be with ominous veiled threats of what kind of fish food I will be if we double cross them. He gets up and leaves after telling me to stay there for 10 minutes before leaving. As he walks down the steps and off into the park, I see Reek emerge from behind a truck, now wearing the same drab coveralls as the workers, and start following the baron.

I tell Gunther that is supposed to be in the car to pick up Reek as soon as he finds out where the baron went to. He says he will as soon as he gets back to the car. I ask him where he is, then notice that he is in the park. Wanker. I tell him to pick me up which he rushes to do, and we soon retrieve Reek who had followed the baron until he got into a car and sped off. We go back to the hotel for a session of after-action and a finger wave with Gunther if he screws up again. He is basically following the Swiss orders not ours, so we treat him like what he is, a Quisling.

I had gone to the Swiss guy's home with Gunther and knew a bit about him and his family. He had a daughter that was a heroin addict, not uncommon in those days as cheap heroin was flooding Europe and hooking the next generation. His wife was a nice woman trapped in a hopeless marriage with a sociopath. The house was a fortress with cameras and full-on security system. Jeff was doing a security survey in case someday we had to come back and do some long-range target practice through the windows facing the cliffs to the east. He had used a sling shot from a nearby copse of trees to launch rocks at the perimeter and as soon as the alarms went off had timed the police time to respond. He was also taking pictures and using his pals at Interpol to get IDs on anyone we met. This was also our insurance policy if the Swiss decided that we were going to be part of the bust to be done by the locals.

The meet is scheduled for three days later at the Ferrari dealer's place of business in St. Gallen. I inform the Banker back in California and he is as antsy as Gunther. We had to delay it a day as one of the gang hopped on a plane and went to Milan. Jeff hopped a ride on the private jet and got there a couple of hours before he arrived and was able to get pictures of who met him. He called after he lost them in traffic, and I told him to come on back but give the info to the boys.

On the appointed day Gunther drives Reek and me to the meeting. We are wired so that he can hear our conversations and he is there to be our over watch. Jeff, who is now staying at a separate hotel, is trailing us and watching him. We are both in business suits and we have some flash money with us. Not enough to be jacked, but enough to whet the gang's appetite. The big money is in escrow for the exchange. I notice that Reek is wearing a very expensive Armani suit. He looks great, except he is wearing Doc Marten steel-toed boots with it. I chastise him and he tells me that before the day is over, I will thank him for that addition to his wardrobe.

We go inside and are ushered into a room with windows facing the parking lot not 30 feet away. Sitting on the U-shaped leather couch are the German baron, the Ferrari dealer, and the scion from Garmisch. We are positioned across from them. The four Russians are sitting on chairs with their backs to the window. We have opted not to go armed. I have a gas pistol, which were allowable at that time, and we both have stilettos if things go really bad, figuring we could get guns in the melee if it came to that.

I am sitting on the corner and Reek is facing the three on the other side. The Russians are staring at his boots occasionally. I look up and out the window and realize that Gunther has parked the car right in front of the window and is hunkered down like he is taking a nap. I want to kill him, but no one has noticed him yet except me.

The baron acts as if he is in charge and starts this as if he has not a worry in the world. I explain to him that Lloyds will follow the procedures as outlined with their client, the Japanese bank. He demands 500,000 Swiss francs and I tell him that we have 25,000 to show that we are interested and serious and pull the escrow documents out of my briefcase, gently rubbing the stiletto to reassure myself. Reek just stares at them and occasionally adds to my assertion that we are going to control the exchange, not them.

I slide the money in an envelope over the table and state that we can conclude the transaction, but it will be on our terms. The baron leans back on the couch with both arms on the backrest and sneers at me, indicating that if he gave the signal we would never leave this room, and glances at the Russians as if to say, and here is my proof. I am getting sick of this pumped-up plastic bad guy. I am about to lay it down on him and lean slightly forward.

A huge gob of what looks like snot and afterbirth the size of a tennis ball shoots past my face and impacts right on the baron's forehead and face, making his startled face sheen in the light. Reek comes off the couch, almost spilling me in the process, grabs the table, flings it aside and grabs the baron by the throat and shoves him into the wall. He is right up close to the baron now, telling him that he is a little pissant and that he personally is fed up with him and his Russian girlfriends. Everyone is looking at Reek now.

The Baron has actually pissed himself, with the stain widening on his tailored slacks and dripping onto his handmade Italian tassel loafers. Reek has him up far enough that his shaking has caused one to fall to the floor. His two buddies are leaning away from the two of them but are now trapped by the overturned table. I look over to the Russians and they are staring also, but they aren't centered on the baron, they are looking horrified at the boots Reek is wearing and wondering if and when he is going to use them.

He drops the baron into a heap and slaps him across the face and tells him to repeat his last statement. The baron hasn't recovered his voice or nuts and just lays there terrified. This brings the meeting to an immediate end. They don't have the paper or at least didn't bring it because they felt they could get at least 500 grand in cash. That's shot, but they know we are at least serious, not two slick lawyers from London town that they could intimidate.

I look out the window and see that Gunther is sitting bolt upright and staring at us through the window. I talk into the wire and tell him to get his ass to the doorway with the motor running, we are coming out. I tell them when and where we will meet, in a public place, where the exchange will be made. We back up carefully and I nod to the Russians as we go out. They are the ones who obviously have control of the document if not actual possession and we are out the door.

We get in and Reek slaps Gunther across the back of the head so hard I thought he knocked him out and hisses the next time he, meaning me, tells you to do something and you don't, I'm going to beat you to death. Gunther believes him—hell, I believed he was going to do it. Once he gets the rage up, there is a bursting radius. I had seen it on a lonely hilltop in Vietnam with him surrounded by enemy dead and wounded. He and his Nungs were compacted into a shell hole. We had come in to retrieve them. He had earned a Distinguished Service Cross that day, they should have arrested him for littering; so I knew what he was capable of. The Russians certainly suspected that if it got dicey, they were going to have to work for it. It just goes to show you that you can dress up a bear, but that doesn't mean you can shave him. We drive back to the hotel and Jeff puts a discreet tail on the group until they break up.

Our exchange is the following day, we now know where all the gang is located and the Swiss is ready at the bank. We follow the two patsies from where they leave their car. It's Fasching time and folks are celebrating the German version of Mardi Gras with their usual abandonment. What the patsies don't see is that Reek is on top of them, he weaves up to them appearing to be just another reveler and he puts his arm around their shoulders and says lets party gentlemen, They turn as if to tell him to bugger off and realize who has them in his grasp. They break free and run for the car park, not for their getaway vehicle, but to get close to a police car that is parked in front of a *schnell imbiss*.

I get to the meeting place and the baron is waiting for me, sitting at a beer barrel being used as a table. He is as nervous as hell and is rubbernecking all the time. I tell him to calm down and slide the envelope with the 25 grand in it across the table and tell him to give me the document. He sputters some bullshit then grabs the envelope and flees. He gets to the top of a flight of stairs leading down and onto the plaza below, when he realizes that the peripheral object hurtling towards him is Reek. He panics, trips and tumbles down the stairs, landing in a heap below. He leaps to his feet and bounds off like a star sprinter. I yell at Reek to get the envelope that he dropped, and he stops, retrieves it and comes back to me. Jeff joins me at the top of the stairs. He remarks to me that when they question the baron, they should find out who his tailor is. He asks if I happened to notice that when the baron got up from his tumble, his clothes looked hardly soiled and had not wrinkled at all from the abuse. I look at him and he adds, that is the reason to buy quality attire.

It is apparent that the baron and the two bozos don't have the document. The Russians have it and have decided this is too much work for 500,000. The Swiss gets a Supreme Court Justice up from bed and has a warrant within the hour. There is a Bolo out for all of them. The threesome are arrested within an hour, and the Russians at their hotel; they are all thrown into the local hoosegow in the worst cells available. The threesome crack within hours and are singing like nightingales. The Russians settle for pragmatism after three days. They had set up home in their new abode and even had some laundry hanging to dry. This must have seemed like upscale from their Russian apartment back in the Motherland. They agree to hand over the document in return for no charges and deportation to their next crime scene.

The document is found, and we take it to the bank. We authenticate it with the rice paper proof and watch as it is locked in a safety deposit box with no back door and we are given the key. We have arranged to have a courier pick it up and deliver it to the Banker. The Banker will then transfer it back to the original owner. All done simultaneously when it hits his bank, at that point the banks are in control. The Swiss get their bite, we are paid our facilitation bonus, and the rice-and-sake crowd get their loose cannon back.

Before we leave Germany, we have to say farewell to our fair assistant and bon vivant. Reek and I take her to the airport. It's a tearful farewell. She won't be able to enter the US for two years because of the stipulations in her declaration of self-deportation. She is rightfully indignant and hurt because of her treatment at the hands of Customs and Immigration. So am I. I ask her if she needs cash, and she gives me that shy, vulnerable look that women get when one suggests a charitable contribution is in the wind. I give her 6,000 Swiss francs in crisp new bills. It disappears in an instant to be sequestered in the pockets of her new wardrobe. Reek looks at me, I look at her, and he says we should shake her down. We are manhandling her and she is fighting us off like she was Jackie Chan. We find another ten grand, but we only manage to search half of the suspected hidey holes. She is fighting us off with feigned indignation.

All during this time when I would give her cash to buy things or pay for bills, I never got change back. It wasn't all that much, a hundred here, a fifty there. I chalked it up to feminine wiles. All of us know that if you are with a woman and if there is a female clerk at the register, you hand over a hundred to pay for something and the clerk always hands the change to the woman. She gets all the loot back because we are laughing so hard. She added the spice and spunk to what had been a great operation. We watch her go through the security point. As she leaves, she turns and gives us both the two-fingers-to-the-eyes Sicilian hand signal for a curse, smiles and walks out of our lives.

We don't even get to New York before Gunther calls and says that there is a problem. He claims that the Secret Service grabbed the instrument from the courier company and that it was a forgery and so was the proof. The Banker isn't taking my calls.

We get home having to have had the Penguini standing by but never used. I am now getting calls from the Banker. He is spouting the same story. I tell him that I don't believe him and that if the US government had seized the document, I would be crowded by gents in polyester. It goes back and forth then he disappears. Bottom line is that we aren't getting our bonus.

It takes a few months, but I have an idea what happened. The safety deposit box hadn't been safe at all, the Swiss and the bank in Switzerland contacted the Japanese directly after they had possession of the document, cut the Banker in the US out and discounted the note to the Japanese for a few million. The Banker was a funny guy when he wasn't trying to cheat you. He was a first-class schmooze and would eventually get involved in something he couldn't talk his way out of. There was no use going after him, it was the Swiss who had screwed everybody, him and his pals in government. As much as I hate to be screwed, the Banker was a loose end for that crowd, we weren't.

I sometimes daydream about lying on that hilltop looking at the windows through a high-powered spotting scope, beside Jeff sliding the bolt forward on a big, precision long gun. Afterwards we would link up with Reek who had been watching our backdoor and then hike over the Alps to Austria, while savoring a pleasant memory, kind of like the last scene in *The Sound of Music.*

CHAPTER SEVEN

Ishtar and Other Bad Plots

In our world, things change because of world events, and they change rapidly. I often feel like the baboon leaping from rock to branch to log, trying to cross crocodile country. You must always have alternatives. Since our European adventure, Jeff has solved the problem of revitalizing the training programs, and sales while I have been down South. He has acquired a new partner and corporate structure. It does basically the same thing we had been doing but added consultation at high levels with foreign governments and our own.

His new partner Cosmo is a more urbane copy of the Badger. Oily and scheming, he wears his MBA like a ribbon display on a uniform. It would seem very likely that he had the proper contacts in a number of countries, as his business adolescence had been spent in the hire of the Aero-prick Defense Industry. I am amazed at the number of high-level people that he actually does know and know fairly well. They range from princes and sheiks to generals of various armies, and intelligence operatives of some of the more astute agencies. Throw in a president or two, and a few underworld figures, and you have his network.

On top of that he has excellent recall and can remember the name of the concierge at the George V or a taxi service in Marseilles that will take you safely to the waterfront. He is personable and people do him favors. He is smart enough to know that the Beast is a good asset and he realizes that Jeff has the memory of a main frame. It irks me that he can recall minutiae and names from our travails, yet cannot remember to bring his wallet when we are drinking. But we all have our faults. I admit that I sometimes rehearse fibs before using them, so that they are believable. It is a development from my misspent youth.

Cosmo has wrangled work from one of the defense companies and he and the Beast are headed for South America. The Andean Pact countries are planning to launch their own communications satellite and Cosmo is trying to help his US client screw their French competitor. The trip starts in Bogota, still in the grip of the Medellin and Cali cartels and locked down like a prison. Jeff needs a haircut from the barbershop half a block down the street from their hotel and the concierge assigns two armed guards with an enormous Rottweiler just to accompany him the 50 yards down and back. They have some kickback time, so Cosmo decides that they are so close they might as well use this time to visit the Amazon. They catch a small, local airline and head down to Leticia, the nail on a slender finger of Colombia that reaches down to the river.

From the start it's apparent that the sterile Wild Kingdom view presented to the American public leaves out a number of things. Mud being high on the list. It presents not only the difficulty of getting through it, but the rank smell of decaying swamp that clings to everywhere.

The next would be critters. Not only do they have the malarial mosquito, but it has cousins in abundance and as soon as dark starts to steal over the land, their numbers go from a few hundred per square meter to swarming host numbers. The basin also has a number of parasites, pests, and bizarre creatures that one thankfully doesn't find in the States.

They check into the Hotel Anaconda, an ancient structure but moderately comfortable. The room's air conditioner sounds like a trans-Atlantic flight on a C-130 and manages only to reduce the outside climate of 98 degrees with 98 percent humidity to a balmy 85 degrees with 98 percent humidity inside. In the hotel lobby they hire the same man that guided Jacques Cousteau up the river for his award-winning documentary on the pink dolphins. The Beast fills me in on the smell and critters that were left out of the travelogue literature.

They are sitting in a bar built on a barge moored at the river's edge, drinking tall, icy bottles of Antarctica beer, which is a surprisingly good, crisp lager. They link up with their guide, a small, wizened, white-haired sexagenarian who actually kind of looks like Jacques Cousteau. He is loading the boat for the trip up-river whilst Cosmo and Jeff play Bwana and relax. The air is heavy with the pressing humidity and it's hot, Amazon hot, which is very much like Africa hot or Vietnam hot. Jungles are jungles.

They happen to notice a stack of Plexiglas sheets about 8 feet long and 3 high with rubber bungee cords and slots that receive the hooks on one side and the ends. Curious, they ask what they are for, and their drinking companion, the bartender, cocks his head toward their guide busily loading the boat and tells them that it's for the cannibals up-river. Now we all have had at least one local Bubba from anywhere try at one time to pull our wanker with stories of Croc-a-lopes and hairy beasts, so Jeff chocks it up to stories they tell tourists. They get their gear and head down the walkway to the dock and then the boat. These are not the handrail-equipped gangplanks you see in the movies; these are single hand-hewn planks covered with slippery mud or wet with the incessant rain.

As the Beast climbs aboard, he notices that beside the shotguns and a high-powered rifle scattered about the boat, there is an old M79 40mm grenade launcher shoved incongruously into the mix. Next to it is a bandoleer of 10 grenades. The guide has two assistants, both Yagua Indians, one of whom seems to be stumbling drunk. The sober Indian carefully maneuvers the boat out into the channel and they are soon moving along with the wind now cooling them off just a little bit.

Jeff engages the guide and asks about the weaponry. The guide explains that there is a tributary running south deep into the Brazilian jungle, where a tribe of cannibals lives. He says that the Plexiglas sheets are erected in the boat as protection from the arrows and blow-darts that they fire from the position of lying on their backs.

The guns and the grenade launcher are there because once you pass their village on your way up-river they know you will be coming back down. The tribe then builds booms made of logs and limbs that stretch across the river, trapping your boat. In that case you have to blow the booms away so that you can get through, hence the M79. Of course, going up that particular tributary is only during the dry season when the main channel gets low and costs extra, so they decide to stay on the main river and head up toward Peru.

As an aside he points to the mouth of a river that empties into the Amazon and says that it is the one he is talking about. As they are looking, a figure steps out of the jungle and views their passing. He is a short but well-muscled man, with no excess fat, built similar to the Yagua tribe they had been traveling to see. Unlike the Yagua who paint their face with red or orange, this man's face was painted with ashes giving him a sinister look.

The Beast notices that he appears to be naked with an erection. The guide laughs and says that it's not an erection but rather the practice of driving a thorn up the penis opening and tying the loop on the end around their waist so that their junk doesn't flop around when moving. He adds that the reason is that there exists an insect that is attracted to the genital area, and the binding rig with thorn stops the attraction factor and also stops the beastie from getting up the tract. The insect's bite is so painful that it is crippling.

Jeff is thinking that it's perfect that Cosmo is along as an entomology entrée and checks to make sure his gaiters are tight, and his fly is zipped. The banks of the river pass lazily by as they move upstream with the guide entertaining them with his past escapades. They stop at a Yagua village, met by a group of very short women all naked above the waist. Their hosts insist that they have their faces painted with an orange paste. This, they believe, will prevent the visitors bringing evil spirits into the village. They tour the village and are taken out into the nearby jungle where their sober guide shows them how to cut a specific vine and drink fresh water out of it. He also tortures his drunk partner by finding a very large, black, and apparently dangerous species of ant and threatening him with it. This sends the drunken Yagua screaming into the jungle. They don't see him again.

From the Yagua village they proceed back down river to the Ticuna village. The Ticuna are a more advanced tribe than the Yagua, who still seem right out of the Stone Age. The Ticuna live in similar huts, but with a few more modern conveniences and they mostly wear western clothing. Chicago Bulls T-shirts seem especially popular amongst the young men. The chief of the village's daughter, a pretty teenager with beautiful, immaculately white teeth, gives a demonstration on how they cook their staple food, a flour made from cassava root.

Continuing down river from the Ticuna village, with Colombia on their left and Peru on their right, the guide suggests they stop for a beer. He pulls the boat up in front of a pair of rickety structures built on stilts above the river on the Peruvian side. One is a single-room general store with a meager supply of staple foods, tools, etc. The other, reached by walking 8 feet across a slippery board, is a cantina of sorts, but they do have cold Antarctica beer. After relaxing for about half an hour, the proprietor comes to the table with a favor to ask; can they wait around a little because someone is coming who wants to meet them. "Is it a cop?" is naturally the first thing out of Cosmo's mouth, but the proprietor assures them it is not.

A little over an hour later an old man walks out of the jungle and climbs up to the cantina. Gnarled but fit, dressed as a local Indian from the waist down in a sarong and sandals but with a T-shirt. They are perplexed as to how he heard about them and what he wanted. The bush telegraph is the answer to the first, and apparently, he wanted to talk to them about his visit to America in the 1930s. He is obviously from one of the local tribes and they are astonished that he visited America as a young man. It turns out that he had been drafted into the Peruvian navy and had been assigned to a destroyer that made a hospitality visit to San Francisco for the World's Fair.

He had wandered around the city on shore leave, amazed at the exposure to an urban setting. In terms of culture shock it must have been the same as if we were transported to some alien planet of an advanced race. It turns out that he hadn't any questions, he just wanted to talk to them in broken Spanish and pidgin English, of his travels. Where you don't have television, storytelling becomes an art form.

* * *

A month or so after his return I am with the Beast for lunch, during one of my occasional visits to their office. He is telling me stories of the Amazon and we are trying to figure out what's next. But as Rick said to Louie in *Casablanca*, "It looks like fate has taken a hand," because the news is suddenly full of Saddam Hussein's invasion of Kuwait. When Saddam invades, Cosmo just happens to be in Saudi Arabia with a retiring Marine Corps general who was on terminal leave (his retirement became official on the day of the invasion). They are with an old friend, Rashid, who works for the Gulf Cooperation Council. Of course, the invasion sends shock waves through the kingdom and everybody is frantic to get the general's reassurance that the US will come to their rescue. The Saudis are terrified of Saddam's supposed stockpile of chemical weapons and everybody in the kingdom wants a gas mask.

Cosmo of course sees this as a great opportunity and calls Jeff to have him look for surplus army gas masks to re-sell in Saudi Arabia at an extreme mark-up. Sadly, once GHW Bush sends the 82nd Airborne the gas mask hysteria will die down along with Cosmo's visions of war profiteering.

But Rashid had mentioned something else to Cosmo. Through his father, who had been run out of Iraq by the Ba'ath party revolution, they had contacts high in the Iraqi military to include the general staff. Rashid's father had been the owner of a bottling plant in Basra and lost everything, so he had no love lost for Saddam. So Cosmo also asks Jeff to find a contact who could take intelligence developed through these contacts on behalf of the US.

Jeff's first stop is SOCOM, where he is shunted to a female major in the G-2 who is completely lost. "Well we don't have any protocol in place to work with civilians, yada, yada, yada." Typical REMF. Asking around though, he gets the name of a lieutenant colonel up at the Pentagon who just might be interested.

Lieutenant Colonel S. is more than excited about the idea, and he is just about to depart the puzzle palace and head for the US Embassy in Riyadh. He does however tell the Beast that it would make his life easier if somebody in the network had a military ID card. So Jeff somewhat reluctantly went over to the 12th Special Forces reserve group at Los Alamitos, CA and re-upped prior to departing for Saudi Arabia. This is akin to laying your junk on a stump, giving your ex-wife a hammer, and thinking to yourself, "I hope this doesn't hurt."

Starting in Riyadh we meet with the Lieutenant Colonel at the embassy (where you could get a beer and drink it out by the pool, absolute Heaven!) and put him in touch with the right Arabs. The Lieutenant Colonel speaks Arabic so fluently that the Arabs are constantly amazed. They would run and get their friends just to hear a gringo speak such perfect, colloquial, unaccented Arabic. We quickly get the intel program up and running. Did it help? Who knows? One only hopes that good deeds will factor in, if they catch you at something heinous.

In the hotel lobby we also run into a Sprint representative who invites us to spend an evening at the *diwaniya* of Sheik Abdelaziz al-Tuwaijri, the commander of the Saudi National Guard. This takes place in a goatskin tent about the size of the average high-school gym. We meet his sons Hamed and Khaled who become friends and will help us out later. The sheik invites us to share a Ramadan breakfast (right after sundown) and sit near the head of the table on his right-hand side, a big honor in that society.

From Riyadh we drive the 240 miles over to Dammam, arriving just in time to see General Schwarzkopf's briefing about how the war was won. Lieutenant Colonel S. is there, he had been the translator for Schwarzkopf during the final cease-fire negotiations. We have Rashid's brother Sefa with us, whose wife had been trapped in Kuwait through the entire occupation, and he is even more anxious to get up to Kuwait than we are. We're going for money, he wants to insure his family's safety.

We have no problem cadging some permits and special passes, but Sefa is another story. We sign him on as our driver in exchange for him giving us his brand-new Lincoln Towncar to use. When we go over to the office where permits for Saudi and Kuwaiti nationals are issued (and at that point they are basically giving out none) it turns out the person in charge had been one of Sefa's professors when he got his engineering degree at the University of Kuwait. "So Sefa, this is what you have become, a driver for the Americans, and you had such promise when you were my student, so sad." But suspecting there is more going on than meets the eye, he gives Sefa a permit anyway.

We head north, our shiny black Towncar looking very strange, alone amongst all the convoys of military vehicles on the road. Most are headed south, long convoys of lowboy trucks carrying M-1 Abrams tanks and every other conceivable military vehicle, but there are still a lot of supply trucks heading north.

When we get to the border the Bedouin border guards are totally non-plussed; this is the first civilian vehicle they have seen, and they have no idea what to do with us. We show them our permits and passes but it doesn't help. Finally, they say our permits have to have the stamp of the prince of Khafaji. They say without this stamp there is no chance the Kuwaitis will pass us through their checkpoint even if the Saudi side let us through. Khafaji is a small town just south of the border, the site of the only battle of the war on Saudi soil.

We arrive in Khafaji and it is a complete ghost town, still showing all the pockmarks of the battle. We eventually find the Prince's palace and wander in, only to be confronted by two very old men. The prince has to be in his nineties, and his "secretary" is probably in his mid-eighties. They are both totally confused by our request of getting our papers stamped. The secretary starts rummaging through stacks of cardboard boxes that litter the floor, mumbling to himself about what an unusual situation this is, until he finally retrieves some kind of a stamp. While he is doing it the Beast espies a medium-sized box jammed full of passports, over a hundred of them from every conceivable country, but he can't figure out how to spirit the box out of the room. This is an old Special Forces maxim: if it's loose, lift it, you never know when it might be useful.

Once the papers are stamped and signed by the prince, we head back to the Saudi border post and the confused guards really have no choice but to let us through. Of course, they had lied: not only is there no Kuwait checkpoint, there isn't even a road. The Iraqi army have torn up the pavement and we have to detour east into the desert and follow a dirt track several miles to where it finally reconnects with intact pavement.

Kuwait City when we arrive is a mess. Every corner has some kind of silly little cinderblock pillbox, there are abandoned T-60 and T-62 tanks everywhere (all facing north with the main gun locked down in travel mode and a full complement of ammunition), strange Arabic graffiti on every building. Sefa is especially upset by one particular piece of graffiti but he doesn't want to tell us what it says. He says it is just too awful to repeat. We keep bugging him until he relents and in a very soft voice, he says, "Your mother and two goats." Okay, it must be a cultural thing.

Thankfully Sefa's wife and her father's family are okay, and really happy to see us. Especially since the trunk of the Towncar is stuffed with canned food. We drop Sefa and a load of food off and head downtown to find a place for ourselves. The only place open (sort of) is the International hotel, a former Hilton next door to the US Embassy. As we approach, stepping gingerly across concertina wire, near the door we spy a small, white car that had been hit by some sort of aerial-delivered munition and blown apart from the inside out. It looks like a flower with sharp, jagged petals.

The hotel is nearly empty, so we sit in the lobby and open some tuna from our stash in the car. It brings a couple of dozen feral cats out of the woodwork who surround us and when we put the empty cans down on the floor, they fight viscously for the right to lick up the remaining juice. Dinner and a floor show!

We spend the night and one of the discoveries we make is that there is no water and the toilets don't flush. Fortunately, there are so many empty rooms that when you fill up a toilet you can just move. They get the water back a few days later.

The few Kuwaitis that were left behind in the city throughout the occupation are incredibly thankful to the Americans. More than once we are approached by men on the street chanting "George Bush, George Bush" and wanting to shake (or in some cases even kiss) our hand. This doesn't last long. When the Kuwaitis that had ridden out the war in Saudi Arabia come back they are much less appreciative and by the time the upper classes who had spent the war in London or the south of France get back they were downright annoyed that the Americans had ended their vacations. The half-life of Kuwaiti appreciation turns out to be about two months.

A day or two later we are up on the Jahra ridge, the so-called highway of death. By the time we arrive the visible bodies have been mostly removed but there are still plenty trapped inside destroyed vehicles and the coppery-sweet smell of death is almost overpowering. Every type of rolling vehicle imaginable is represented. Cars, trucks, construction equipment, if it can move under its own power an Iraqi is stealing it and trying to get out of the country. The A-10s had had a field day. My most vivid memory is a stake-bed truck that had headed off road to the east and been hit dead center by a CBU. The truck had been full of boxes of women's shoes looted from some boutique in Kuwait City and the shoes were blown in every direction. The destroyed truck sat there in the desert surrounded by concentric circles of shoes of every imaginable color. It was actually quite a work of art.

Our mission to help rebuild this country finally gets started in earnest. The first order of business is to put out oil-well fires and then find a way to clean up millions of gallons of crude oil that has flowed out onto the ground. There are literally lakes of oil, hundreds, maybe thousands of square meters in size. We wander into the US Embassy, which has just reawakened, to see what kind of help we might find.

Incredibly we walk right into two members of the 10th Special Forces Group from Fort Devens, old friends. They are astounded: "What the hell are you doing here?" "Just tourists," we answered, "we heard this was the hot vacation spot this year." The great thing was they have a BBQ set up and are cooking burgers. After a week of canned tuna and canned string beans a burger is absolute ecstasy!

Another interesting food-related event happens a couple of days later. We hear through the grapevine that the Holiday Inn down by the airport is opening its restaurant. Real food! We drive down and enter the restaurant to find a full mezza, tabbouleh salad, falafel, various Greek and Lebanese meat and vegetable dishes, paradise. We stuff ourselves right up to closing time.

The Holiday Inn sits on a bluff above the sixth ring road, which is a limited access freeway. Near the hotel is a long, downhill, curved access ramp to the freeway. At the bottom of the ramp at the point where it joins the main highway the Kuwaitis have built a sandbag bunker which is manned by a single militiaman who is waving his arms to flag us down. Miller is driving so he pulls up next to the bunker and in an excited voice the Kuwaiti screams, "Don't go, fire!" "What?" asks Miller and the Kuwaiti repeats, "Don't go, fire." Sefa has dozed off in the backseat so we shake him awake to ask what the hell the Kuwaiti is talking about, but before he can ask, it became clear. From across the highway a string of tracer fire starts hitting the bunker and ricocheting off the pavement right in front of the car. Miller slams the transmission into reverse and peels out up the onramp. We haven't traveled 50 feet when we come upon the first in a convoy of Egyptian army Mini-Mog trucks coming down the on-ramp in nice convoy formation. Miller passes the first truck on the left as the driver slams on the brakes and stares wide-eyed at this nearly out-of-control Towncar speeding by in reverse. Because of the curve of the on-ramp Miller has to slalom, passing some trucks on the right, some on the left until he reaches the top of the on-ramp, does a J-turn and drives around behind the Holiday Inn out of the line of fire. We sit there hyperventilating for about 15 minutes until the firing stops and then restart our journey back to the hotel on back roads. By some miracle there is not a single bullet hole in the car and not a single collision, even a minor one, with an Egyptian truck. Sefa attributes this to Allah, we attribute it to all that driver training we have done over the preceding nine years!

A couple of days later there is another tracer fire incident; there are obviously still a few Iraqi stay-behind units or individuals causing trouble. This time Sefa is driving when tracer starts crossing the highway a couple of hundred yards ahead. We tell Sefa to pull over, but he keeps driving and calmly says, "Allah will catch the bullets in his hand." We tell him Allah might be busy doing something else right then and to PULL THE GODDAMN CAR OVER. He does.

Cosmo goes back to the States and we move into a villa across the street from the SAS Hotel down south of the city. It's a nice place, all marble with about six bedrooms for visitors, but come summer the air conditioner can't cool the place lower than about 85 degrees on a good day. Of course, that is better than the normal 120-plus degrees outside, sometimes with high humidity! We hire a secretary/translator, an Egyptian girl who speaks fluent English and French as well as her native Arabic in a variety of dialects. We also hire a driver with his car (Sefa had returned to Saudi Arabia with the Towncar). Our driver is a young Palestinian man who had been an English teacher before the war but couldn't get his job back because of the irrational hatred of Palestinians by Kuwaitis right after the war. True, about two-thirds of the Palestinian population had sided with Saddam, but at least a third stayed loyal to Kuwait, including our new driver. Sadly, all of them got tarred with the same brush.

One day we head out to the airport to pick up a couple of visitors from the States. Just outside the airport is a checkpoint manned by the Kuwaiti militia, whose effusive gratitude for being saved by America has by now morphed into a simmering contempt. When the militiamen see the driver's Palestinian ID they decide to give us the full treatment. Everyone and everything out of the car; they pull the seats out, pull the floor linings and the trunk liner out, throw everything down into the thick dust on the ground. The Beast is working his way into a nice rage, but our secretary Maria is standing next to him, coaching him to keep his cool: "If you do anything you will only make it worse," etc. Finally, they allow the driver to reassemble the car and we proceed to the airport. But they keep the driver's ID and say we will have to pick it up on the way out.

We pick up the two suits who have flown in from D.C. and their luggage and head out of the airport, but of course we have to pull into the checkpoint to retrieve the driver's ID. Well, the full treatment starts all over again, including throwing our visitors' luggage into the dust and pawing through it. Too late for warnings from Maria, the Beast loses it. He heads over to the five-foot-six corporal in charge and sticks a finger in his face and tells him, "Enough is enough, put everything back and we're getting the fuck out of here right now." Well the corporal starts yelling into the Beast's chest in Arabic and before anybody realizes what is happening Jeff has punched him in the face and knocked him on his ass.

About 30 yards away is a GP Medium tent (US Army issue) that appears to be headquarters for this circus, so the Beast grabs a fistful of the corporal's BDU shirt (US Army issue) and drags him over and flings him through the open front flap of the tent. Upon entering he finds a Kuwaiti captain sitting at a field table (US Army issue) with a shocked look on his face as the corporal's semi-conscious body rolls to a stop in front of his desk. Miller pulls his military ID and slaps it down hard on the desk, loudly proclaiming, "I'm Sergeant Jeffrey Miller, US Army Special Forces, and this man has insulted me and my guests and I want to know what you're going to do about it!"

Now it is Jeff's turn to be taken aback, as the captain leaps to his feet, comes to rigid attention while throwing a palm-forward British-style salute and yells at the top of his voice, "Fort Bragg North Carolina!" He continues: "I trained with the

US Special Forces at Fort Bragg, I love the US Special Forces." The captain comes around the desk, yelling Arabic invective at the corporal who is just sitting up. This is punctuated by the occasional kick. The captain pulls him to his feet and frog marches him out to our car where he gathers up the other two militiamen and supervises them dusting, cleaning and reassembling everything, all the while screaming at them in Arabic. It is a world-class ass-chewing, even without understanding the words.

Jeff of course sidles over to Maria and the driver and is saying sotto voce, "Don't say anything huh, gonna make things worse huh, stay cool huh." Once the car is all put back together, we saddle up and head back to the villa.

* * *

There are some interesting characters in Kuwait these days, and one of our favorites is a Brit we call Lawrence of Arabia. His name really is Lawrence and he sells kitchen supplies. He drives all over Kuwait, Saudi Arabia, and Iraq in an old Yugo that just barely hangs together. It features a rag as a gas cap and a unique three-stage braking system. When Lawrence wants to stop he pumps the footbrake hard, bleeding off a little speed, and then locks up the emergency hand-brake, slowing the car some more. Finally, he sticks his left leg out and uses his foot to bring the car to a complete stop. It is terrifying if you are anywhere in his trajectory, but hilarious if you are sitting up on the balcony watching him pull into the parking lot. Later we heard Lawrence might actually have been an MI-6 agent which made sense. He was just the kind of cultured, eccentric Brit that John le Carré would feature in one of his books.

Another favorite is José, scion of one of Argentina's richest and most powerful families. Jeff is sitting in the lobby of the International Hotel one day, drinking coffee and smoking a Cohiba cigar from a box that has just been presented to him by a visitor from Bahrain. José walks by and noticing the cigar, says, "Excuse me is that a Cohiba?" Jeff replies "Yes, it is," and Jose says "Well of course they are the best." So Miller pulls another Cohiba out of his pocket and offers it to José, who accepts it graciously and from that point forward we become fast friends.

Kuwait of course is dry, alcohol free, that is until you figure out the black market. We ultimately find two sources, one that can supply three bottles at a time of Johnny Walker Red Label and one that can supply one bottle at a time of Johnny Walker Black. José has a contact that can supply vodka and red wine. Between us we throw some good parties at the villa. We head up on the roof after dark, play music at nearly the decibel level of the calls to prayer from the mosque next door, and get sloshed.

One night José leaves about midnight to drive home but gets lost in the alcohol fog and makes a wrong turn into a Kuwaiti military base. One of the guards is a little antsy and puts a three-round burst across José's bow. Well, the next day we have never seen a more indignant person in our lives. He is fuming that anybody had the temerity to fire in his direction. Not only does he talk about it constantly but 10 years later Jeff will meet him in Buenos Aires and find he still hasn't gotten over it.

* * *

Shortly after he got back to the states, I was having a drink with the Beast in Newport Beach and asked him what he thought about his time in the Mid-East. He gets that dreamy look that makes him look professorial or that he is viewing old porno movies in his head as background noise. With the Beast you will get a long but detailed answer, backed by facts and addendums. I signal the waitress and order us both a beer.

He starts in with the Arab culture and its relation to Islam and its most destructive precepts. Our conversation is centered on use of resources and distribution of the same. He points out that he thinks the single most destructive social aspect of Middle Eastern societies is polygamy. He explains that this forces the seething rage that infects these societies. Since only the wealthy and elite can afford multiple wives, this means that 25% of the males have all the women, and the rest are frustrated to the point of fanaticism and rage. Additionally, the male is supreme. Mothers are not allowed to discipline their male children and the males won't because they had grown up pampered and doted on by their fathers and grandfathers. Thus, many males grow up spoiled, vindictive, and petty. That cascades down through their society as a small percentage have all the resources and all the women, and the others are given a manual on how to act, how to treat with your neighbors and most of all how to interact with those not of the faith. Add to that Jihad to conquer and subjugate, thus increasing the supply of women, in the hope that they can finally get laid or even better gaining the martyr his place after death in the very society they are prohibited from enjoying on Earth.

He is finishing his presentation when the beers arrive. I tell the waitress to bring us two shots of tequila, as it looks like he is just getting started. The Beast is essentially a data sump—everything he sees, hears, smells, or gets overrun by is fed into his brain, then he archives it, to later regurgitates his slant on the whole experience. We spend the rest of the afternoon exchanging stories of the burning sands.

CHAPTER EIGHT

Tiny Hostages in the Maghreb

One sunny June day Jeff answers the phone in his kitchen—those were the days when phones stayed in the kitchen where they belonged and didn't follow you around everywhere—and a deep, stentorian voice asks, "Is this Jeffrey Miller?"

Jeff tentatively answers "Yes."

The voice continues, "Have you been to Algeria recently?"

Okay, now he knows what this is about. "Well, you know I have, or you wouldn't be calling me would you?"

The voice continues, "This is serious Mr. Miller, the Algerians are not happy with your actions."

"Well we didn't think they would be," replies Jeff.

"A warrant for your arrest for kidnapping has been issued and forwarded to INTERPOL so you had best be cautious with any future foreign travel, don't go anywhere we can't protect you."

Jeff is a little amazed by this. "There are places I could go where you would protect me?"

The voice closes by saying, "Look Mr. Miller, there may be people here who would applaud what you did, there may even be people here who would think what you did was heroic, but officially the US government deplores these kinds of activities by private citizens so just be careful in the future."

At that he hangs up.

* * *

Eight months earlier, just two months after returning from Kuwait there was not a whiff of a potential project in the air. It was starting to approach desperation time. And then on a typically warm autumn day in Southern California Jeff got a fortuitous call from a close friend who worked as a freelance journalist. He told a story about a private investigator up in northern California who was working with a family in the Midwest, trying to retrieve two children who had been abducted to

Algeria, and wanted to know if we could help. Well, desperate times call for desperate measures so naturally the Beast said, "Sure, we do this sort of thing all the time." The reality was that we had only done this sort of thing once before, but what the hell, nothing ventured nothing gained.

Jeff was immediately faced with a quandary of options. I was off in Mexico looking around, while Penguini was self-testing a new set of identification documents obtained through questionable channels, living outside the country under his new assumed name and completely off the grid. He had to find help elsewhere, so he turned to Cosmo. The good news was that Cosmo was intimately familiar with Algeria, having sold aircraft to that country during a previous life, and he spoke conversational French. The bad news was he had no military background and was completely unfamiliar with these types of operations. Jeff decided to call him anyway and he agreed immediately to involve himself. So they placed the call to the family and made arrangements to get together and discuss the project in detail.

Jeff and Cosmo left balmy California for the bitter cold of the upper Midwest. The children's grandparents met them at the airport and took them to their home where they met their daughter Nancy, the mother of the abducted children. That day they spent 10 straight hours hearing the detailed story of how her children were taken. Unfortunately, it is all too common, and we have heard almost the same story many times since. While at school in Europe, Nancy had met and married a young man from an Islamic country. She lived there with him for several years and then moved back to the US. They had two children, a girl and a boy, at that time five and three, and after a few years of rocky marriage had gotten a divorce. The judge allowed the father unsupervised visitation and the first time he got the children alone he just got on a plane and skipped. The next thing the family heard was a rather profane message on Nancy's answering machine telling her she would never see her children again. After a year of giving most of the family's money away to shyster lawyers who accomplished absolutely nothing, they were desperate to try a more direct approach.

Jeff and Cosmo laid out what would become our standard method of operation in cases like this, which involves a fixed-price reconnaissance to establish the viability of the operation and determine the budget. The family agreed to go forward and armed with the address of where the children were thought to be staying, and details about the lifestyle and habits of the father and several of his family members, they set off to do an initial recon and assessment. First stop La Belle France, to try to locate the husband's brother.

Arriving in Paris they made arrangements through a friend of Cosmo to stay in her summerhouse in Normandy, which was located close to where the brother was thought to be living. At this time they were considering the possibility of using the brother in some sort of manipulation to draw the husband and children out of North

Africa and into Europe where they could exert more control over the situation. It seemed like a good idea at the time, but the best-laid plans …

Nancy's recollection was that the brother lived in a "rose-colored apartment building" directly behind a well-known chain store in the city of Caen. They found the chain store with no problem, only to discover that it was ringed by dozens of apartment buildings, all of which were rose-colored, with well over 300 apartment units in total. To top the day off, an icy cold rain was falling, being blown by 30-mile-per-hour winds. Well, into the breach, the only course of action was to canvas the entire area, looking for the name of either the brother or his wife on a mailbox.

Three hours later, having checked between 150 and 200 mailboxes each, soaking wet, wracked with chilblains, they reached the conclusion that the brother had moved, was using an alias, or just didn't bother to put his name on the mailbox. Whatever the reason, they couldn't find him and had better consider other ways to proceed. So they returned to Paris, dropped off the key to the Normandy house, and caught the TGV to Marseilles, home of the largest Algerian community in Europe.

Upon arriving in Marseilles, they checked into one of our favorite places, the Hotel Le Rhul, situated on the Corniche John F. Kennedy overlooking the harbor and the Chateau d'If. The next day they spent all day wandering the Arab quarter with a vague idea of making contact with some sort of underworld or dissident group who could provide help on the ground in Africa. But all they got for their trouble was a lot of strong coffee, approaches from some of the ugliest women imaginable, and a whole new catalogue of strange, interesting, and sometimes horrifying aromas.

Okay, the usual stuff isn't working, now what? Cosmo had the idea of finding a local P.I. to help out, and since he spoke French, he went back to the hotel to check the yellow pages. He called several P.I.s in Marseilles, but most of them had no African connections. Finally, they found one who claimed to have the right connections and invited him and his companion to lunch. This started a series of luncheons in the Hotel Le Rhul dining room with an amazing cast of characters that even Hollywood could never imagine.

First they met Jimmy, who was a specialist in retrieving stolen yachts. After a brief conversation it became apparent that Jimmy was in the business of stealing yachts from the French Riviera, taking them to South Africa, and then getting paid to "find" them for their owners. He just did not seem to have the skill sets they were really looking for.

Even more interesting was "Cut Finger Joe" a handsome, tan, Paul Newman lookalike, alleged member of the Union Corse, who had allegedly committed both the world's most lucrative bank robbery and the world's most lucrative kidnapping. Joe volunteered what he thought was a very important bit of information: "If you cut the finger off your victim to send it to a relative with your demand for the ransom, make sure to send it Fed-Ex, not through the regular French mail service." Getting the idea that Joe didn't really understand what they were doing there they thanked him profusely and kept looking.

Finally, over yet another expensive bouillabaisse lunch in the Hotel Le Rhul dining room, they hit paydirt. A private eye who specialized in the Marseille fish market introduced them to exactly the right person, a local "businessman" who he knew very well through the local Masonic Lodge. He turned out to be one of the great characters of all time, a six-foot-four, 300-pound, bearded and scarred denizen of the waterfronts of southern Europe known as Carlos. Although to this day we have no idea what Carlos really did for a living, he introduced Jeff and Cosmo to an incredible array of friends like Louie the Snake, a skinny, sinister individual with a souped-up motorcycle, and an array of "cousins" in Algeria. To a man they were helpful, honest, dependable, and never asked for a penny!

With the help of Carlos and his network of cousins across the Mediterranean, they prepared to go get a first-hand look at the place. Unfortunately, the very day they were scheduled to leave was the day the Islamic Fundamentalist Party won the Algerian elections, and the military seized control of the government. Not knowing what was happening, they decided to wait a few days in Marseilles for the dust to settle, with Carlos and his people keeping them updated on events. During this period of forced inactivity, they became friends with the proprietor of a small pizzeria about a kilometer west of the hotel on the Corniche John F. Kennedy. Turned out he was an old Congo mercenary from the early sixties and had fought with "Mad Mike" Hoare and others. He also made some of the world's best pizza. After about four days it appeared to be calming down enough to venture across to Algeria, so they caught a flight and leapt into the lion's mouth.

When they landed, BMPs (Russian-made armored personnel carriers) were parked everywhere. Soldiers were in evidence at every corner, and after they rented a car they were stopped at several checkpoints on the way to the only decent hotel in town. Fortunately, the soldiers were mostly concerned about internal threats and a US passport caused little stir. They checked into the Hotel al Dzezira, formerly the St. George which had served as Eisenhower's headquarters for a time during World War II, and studied the maps, preparing for the following day's trip to the village of Lakhdaria where the children were supposed to be staying. In the morning they set out, feeling their way along the unfamiliar roads, using a combination of the map and Nancy's directions to find the village. They made it to Lakhdaria with no problem, but had a little more difficulty finding the house itself, and while they were looking around town they made a very disturbing observation.

There were absolutely no foreigners. None! And by the look of the place there never had been and never would be. If they lingered too long they were sure to stand out like a couple of sore thumbs. And worse, in a tiny place like this everyone probably knew what everyone else was doing. If they were noticed too much, or identified as Americans, the word would almost certainly get back to the father and he could get spooked and move the kids. So once they found the house they drove by without stopping, taking a few pictures from below the level of the car's window so they would not be seen. Then they got out of town fast! Once out of town, they concentrated on looking at every possible route of egress. Our little bit of experience had taught us that getting the children away from their abductor is the easy part; getting them out of the country is the hard part, especially a country as tightly shuttered as this one.

They spent the day cruising the back streets and finding routes that avoided all major highways, as the major highways were riddled with police and military checkpoints. They chanced upon some very small, winding routes that led toward the coast, which were almost untraveled and free of checkpoints. By now they had pretty much decided to infiltrate and exfiltrate by sea, since the village was over 400 kilometers from any border but only 25 kilometers from the coast. Once they found the right route, they had to find a landing site which could be entered and departed undetected. This is a lot harder than most people might think.

In order to infiltrate by sea, you need to leave a mother ship 12 miles offshore, then make a 12-mile run through pitch-dark waters, hitting the shore at a precise point. The landing beach has to be completely unpopulated, but somehow marked in a way that will be visible far out to sea. And hopefully, whatever landmark is used would appear on a nautical chart, allowing the mother ship to locate itself at the precise coordinates. A tall order to say the least. But in one of those serendipitous events that sometimes happen, they found the perfect spot very quickly.

Just a few kilometers east of the point where the egress route hit the coast was a huge power generating plant. With two tall smokestacks lit up like Christmas trees at night, that would be visible for miles out to sea, it was also on the local navigational charts of that part of the Mediterranean. About a kilometer west of the plant a river flowed into the sea, forming an effective barrier for any security patrols from the plant itself. Three hundred meters west of the river, a dirt road went to the beach, about half a mile from the highway and screened by thick pines. Any good coxswain could use the power plant as a beacon and navigate perfectly on to that beach, even in the dark. Now all they needed was a ship.

Leaving the continent of Africa, they returned to Marseilles and proceeded down the coast to Italy, to visit a large shipbuilding operation that Cosmo had done work for in the past. While that location had no craft that would fit their needs, they were introduced to a Polish marine architect who had jumped ship in the south of France and had a history of unorthodox enterprises. The Pole said he knew a German captain in Cap d'Antibes who had just the ship they were looking for, and they could meet him the next day.

So they headed back up the coast to Antibes, checked into a hotel, and went to dinner with their new friend. The following morning he took them to a yacht brokerage nearby and introduced them to the captain and his British partner. They were shown pictures of the ship and given a detailed description of its capabilities. It was perfect! A marine research vessel with nearly unlimited range, it had done research off the North African coast in the past so it would be familiar to coastal patrols. The following day they went to view the ship, took pictures for the family, and made a deal. The ship would cost 50,000 US dollars for the mission, expensive but not unreasonable considering the risk factor. They told the captain that they would present the plan and budget to the family and get back to him, and caught the next available flight back to the States.

Jeff thought they had a very good plan. Little did he know that over the next couple of months every part of it would go bad, and we would all wind up operating by the seat of our pants.

* * *

They returned to California in early December, with the first order of business to recruit a team. With both Penguini and me still gone, Jeff had to improvise. He wound up recruiting from the local Special Forces reserve unit, guys who were highly recommended but unproven. The team consisted of the three-man entry team led by Jeff, Cosmo who would function as beachmaster, and the journalist who originally brought the mission as coxswain of the infiltration craft. Fortunately, he was also a lifeguard, EMT, dory racer, and surfer.

Nancy flew out to California for planning and rehearsals. They rented a hotel suite, so Nancy could stay there, and they would have the living room area for meetings. She was told to bring everything in the way of pictures, records, and documents that might be of help during the planning sessions, and Jeff of course had his pictures from the recon developed.

When Nancy arrived, they first briefed her on the budget. She talked to her parents and together they decided it was worth going ahead and they could raise the money. Then they got down to the serious and complex business of making detailed plans. Many days were spent in that room, drawing detailed diagrams of the inside of the house based on Nancy's recollections and on both sets of pictures. These were so closely detailed that they knew every door lock, window, which way the doors swung in or out, and reams of additional minutiae. Or at least they thought they did.

Jeff drew maps of the village and familiarized everyone on the team with all the various routes. He posted large-scale maps and nautical charts on posters for everyone to memorize the checkpoints, landmarks, and driving times between each point. Pictures were keyed to numbered points so team members could recognize landmarks that they would be seeing for the first time. Once everyone on the team was satisfied that the planning was as complete as it could be, they rehearsed.

The most important part of their rehearsals was the small boat movement. They had to be sure that the coxswain could find the right spot on the beach and more importantly find his way back to the ship at night. Toward this end they rented a Zodiac and prevailed upon a close friend to give them the use of his 52-foot sport fisherman. Naturally, on the day they had arranged for the rehearsals a huge winter storm blew into southern California and they had to contend with high winds, rain, big swells, and chop. The rehearsals went well from a technical point of view. They were able to leave the ship at night and make the 12-mile run into the beach and then find the darkened ship again with no problem.

Unfortunately, when they were practicing beach landings during the day, they did have an encounter with the local sheriff. Apparently it's illegal to beach a boat on southern California beaches without a permit. They told the sheriff they were scouting locations for an upcoming action adventure movie which would star Sylvester Stallone and he very nicely told them how to obtain the proper permits (an advantage of being so close to Hollywood). Over the following days they rehearsed entering the house and other aspects of the mission.

Ready, and just in the nick of time, because the Muslim holy month of Ramadan was due to begin that year in late February. During this month, because of the requirement to fast from sunup to sundown, many Muslims change their sleeping habits and are up all night. The team didn't want to hit the house at 2a.m. and find a party in progress so Jeff felt they had to get in before Ramadan started. Looking back now this was a mistake, generated mainly by the family's urgent desire to have the children back. While the emotion is understandable, Jeff should not have allowed it to color his tactical decisions. Waiting until after Ramadan would have been a much wiser course of action.

Cosmo, who had been designated as beachmaster, departed for Europe to make the last-minute arrangements and flew into a shit-storm of problems. First and foremost, the perfect ship was suddenly unavailable. Cosmo was told that it had been put up in dry-dock to have its transmissions rebuilt, but the brokerage had found a replacement for the same price. This turned out to be a 140-foot luxury yacht, which had neither the range, nor the staying power, nor the cover for being off the North African coast that the original ship had. But they were stuck with it so we would have to make it work or indefinitely delay the mission.

Next Cosmo and Carlos proceeded to Tunisia to meet with the agent in place that Carlos had selected. As related earlier, Carlos is a man of extremes in proportions and appearance, and for this trip he selected an all-black ensemble: shirt, slacks, tie and sunglasses, with a silver lamé sport coat. As would be expected, he immediately attracted the attention of the local customs authorities. Imagine a nice fat tabby cat thrown into the holding pen for dogs scheduled for the gas nap the next morning. To top it off, he was carrying a briefcase full of pamphlets and catalogues of military weapons. Somehow, Cosmo was able to negotiate them both out of the airport, and the meeting with the agent went off as planned. It was more likely the gendarmes had just decided that putting him in handcuffs was going to be a Herculean task.

The agent was told to keep a full-time watch on the house and report any movement back immediately. The greatest fear at that time was arriving at the house and not finding the children there. Although the agent was a local, he was a Kabylie, a member of an ethnic minority many of whom have light hair and blue eyes. He was reluctant to spend much time in the village, since it was a stronghold of Arab Islamic fundamentalists and he was afraid of drawing attention to himself. Cosmo had to drive him to the outskirts of the village and physically throw him out of the car because they absolutely had to have 24-hour surveillance prior to the mission. As he despondently walked toward the village, the agent turned and plaintively said, "But I have blue eyes."

Cosmo and Carlos flew back to France and Cosmo purchased a Zodiac through the same agency that had chartered them the ship. With all the pieces in place, it was now time to go for it. The plan was to have the coxswain meet the ship in France, load and inspect the Zodiac and accompany the ship to Palma, Mallorca. When the ship departed from France, Cosmo would fly to Algeria and handle that end. The entry team, along with Nancy, would rendezvous with the ship in Palma and depart to the dropoff point near the Algerian coast. But when Jeff got to Palma, he was immediately smacked in the face with an array of problems.

Cosmo and the coxswain had trouble linking up in France, and by the time they got together the coxswain had just enough time to get to the ship before its departure. There had not been time to do a thorough inspection of the Zodiac before it was loaded aboard, and upon arrival in Palma it was discovered that the steering mechanism was frozen solid. To aggravate matters further, it was a holiday that day in Spain and they could not get any spare parts. So, Jeff's first act upon arrival was to postpone the mission by one day.

Meanwhile in Algeria, every night Cosmo and one or more of Carlos' cousins would go to the house and practice putting the dog to sleep. At first Cosmo was worried that they might overdose the dog and kill it, but after a couple of days the dog was waiting by the gate for his midnight snack and sleeping potion. During one of these trips the agent's friend Mohammed, an ex-cop, tried to simplify the problem for Cosmo. "Yachts, Zodiacs, sleeping powder, satellite phones, you Americans are so complicated. Why don't I just kill him and then she can marry me? Then I will go to America and eat hamburgers and watch television, and sleep till noon." Cosmo thanked him for his input but decided to continue on with the original plan.

The next day was spent in a flurry of activity trying to fix the steering gear in the Zodiac. They had a specific time that the ship would have to leave Palma in order to make it to the dropoff point at the correct hour and that time was fast approaching. Once they reached it, they had to postpone for another day. When the Zodiac was finally fixed, Jeff had the coxswain and the other two members of the entry team take a 12-mile practice run out into the Mediterranean to test the boat and the water conditions. When they returned, he could see that they were not very happy. The water was very, very cold in February, and the Med had a short, sharp chop that made it impossible to keep dry. Even in their wetsuits, the guys were exhibiting signs of hypothermia, and the team were all quite worried about the effects of the ride on the children.

Now to complicate matters further, the owner of the ship brought Jeff a weather report showing a huge storm blowing in off the Atlantic. According to him there were already 12-foot swells in the Straits of Gibraltar, and if the storm continued on its present track it would arrive at their dropoff point at the same time that ship would. Now he was faced with a terrible dilemma. The money had been spent, everything was prepared, and the assets were in place. If they took the chance and tried to beat the storm, they might be unable to get ashore. They couldn't postpone until the storm blew over, because Ramadan was due to start in three days. They had to either leave that evening or wait for another month. No one else was going to make the decision, it was all on Jeff's shoulders.

All that day he agonized over the decision, going back and forth between going for it and being cautious. That evening, shortly before the last possible departure time, he had everyone meet in the dining room to discuss the situation. The owner of the ship influenced the ultimate decision the most. He said that since they had not left the Mallorcan dock, he did not consider the charter to have started, and they could come back at a later date and use the ship again for no additional charge. The people on the team also said that they would return, having not yet earned the money they were paid. So the decision was made to terminate the mission and try later. It was one of the hardest decisions of Miller's life, but in the long run it was probably the right one.

Jeff placed a call to Cosmo in Algeria to tell him what was happening. He was stunned by the decision, having just been to the beach and looked out over "pond-like water." Jeff told him to catch the next available flight out, and we would see him back in Los Angeles and explain everything in detail. The team and Nancy flew out through Madrid and back to the States. Jeff briefed her parents on the decision and what influenced it. To his surprise they took the news quite well, and he immediately started planning a return trip in April. Little did he know what would be involved the next time.

CHAPTER NINE

Running for Our Lives Across the Rim of Africa

After the first attempt to complete the Algerian mission had to be aborted, Jeff spent the next month making our plan to return in late March or early April. Unfortunately, when he contacted the ship owner with our new schedule he backed out of his original deal. On top of that, he refused to return any portion of the 50,000-dollar charter fee. After several angry phone calls and faxes between us and the owner and the broker, it became obvious that we had been scammed. Jeff started to doubt if the weather report he had shown was real, or if he ever had any intention of going through with the mission at all. There was nothing we could do about it then; we had hoped to extract some kind of revenge at a later time, but heard about a year later that the owner had committed suicide, and his yacht burned to the waterline. I suspect that the Union Corse, which had accepted us into their arms, decided that it was an insult to their honor and that he was a blight on their community that was no longer needed.

Now Jeff was faced with the prospect of no ship, and no way of getting the family to pay another 50 grand for a replacement. First, he knew they didn't have it, and second he felt responsible for the loss. The only thing to do was to change the entire plan to a land exfiltration. This would greatly increase the risk factors for the team members, as well as requiring them to spend up to two weeks on the ground when they had originally been paid for only one night's work. But much to their credit, they both agreed to go back. He had to change the complexion of the team slightly; there was no longer a need for a coxswain, so Jeff let him go and added two more men to the assault team. It happened that Penguini and I were both back in town, so we foolishly accepted the same one day's pay as the others.

In order to save costs, we could not afford a second reconnaissance trip; we would have to combine everything into a single throw of the die. So we packed up and flew to Tunisia to start the operation. The first thing we had to do was find a place to cross the border from Algeria. We decided to split up: I went with Jeff, Cosmo, and the Penguini south, and Nancy headed north with the two original members of the entry team, Gary and Blaine. We agreed to meet and consolidate our information in two days. The four of us headed down into the Sahara, staying close to the border and checking every possible crossing point and smuggler's route. We found two likely possibilities and started back north.

Heading north through a pitch-black desert night, our rented car suddenly blew out both tires on the right side at once, it sounded like twin pistol shots. Afraid it might be an ambush, we kept going for about a mile before we pulled over to inspect the damage. Both tires were completely shredded. With only one spare we were in a bit of a fix. Jeff decided we would put the spare on the front right, since it was a front-wheel drive car, and drove on dragging the right rear rim. The car thumped along until the remnants of the tire were gone completely and then the rim started shooting an impressive shower of sparks into the African night.

After about 10 or 12 miles we came upon a police checkpoint and pulled the car off the road. The checkpoint was positioned in front of a small mud-brick coffee shop with a sand parking lot. Getting out to inspect the rim we found it ground down to about the diameter of a DVD. Cosmo went over and spoke to the cop in French and he eventually flagged down a pickup truck to give us a ride into Kasserine. Because Cosmo could communicate with the driver, he hopped in front. The rest of us hopped up in the pickup bed, only to discover too late that it was about 8 inches deep in goat shit. On top of everything else the night was freezing, we couldn't get down out of the wind without getting covered in goat shit so we had to stand up in the freezing wind all the way.

When we arrived at the Hotel Kasserine Cosmo jumped out of the heated cab with a big smile, "Well that wasn't so bad was it?" We stood shivering, frozen nearly stiff, but eventually climbed down from the pickup only to discover a huge block of goat shit had frozen solid to each foot. We walked stiff-legged like a chorus line of Frankenstein's monsters across the parking lot and into the hotel lobby: the well heated lobby, the lobby with a white marble floor … you can imagine the rest!

The next day, after buying a new tire and rim, we returned to our rendezvous point back in Tunis a day late. Much to our surprise, Nancy and the other two team members were not there. While we were concerned, Cosmo was beside himself. His natural paranoia, never far below the surface, bubbled up like a Yellowstone geyser. He started dreaming up all sorts of implausible scenarios to explain their absence; it became a case study in paranoia. "They went to do the mission without us, it was the plan from the beginning, ever since Mallorca it was all a set-up to leave us behind." Psycho talk! Fortunately, they arrived later that night, having spent extra time to thoroughly recon a particularly good crossing point, the one we wound up using.

Now we were ready to head across the border, so we found a taxi stand where cabs from Annaba, Algeria waited to take people the approximately 135-mile trip into Algeria. We got in two cabs, and the one carrying Nancy and the other two team members immediately turned the wrong way down a one-way street. This caught the eye of a local cop, and the next we knew half of our team was in the local police station. They released the two guys almost immediately but kept Nancy for over an hour. Cosmo of course went into another fit of paranoid overdrive, but our fears of what this might mean were all groundless. It turned out the cop who had originally pulled them over was just trying to get a date. Once he released her, we got back in the cabs and took off.

The cab we were in was an interesting piece of machinery. About every 20 miles the engine just quit, and the cab coasted to a stop. The driver took a bunch of wet rags from the trunk, layered them on top of the engine, then after about five minutes gave the engine a few good whacks with a hammer and started it back up.

The second time this happened we wound up coasting to a stop right across the street from a small brick building with a sign, "Café Palestine." We decided we could all use a piss so we tramped across the street and went in the front door. Sudden silence. Turned out Café Palestine wasn't just a clever name; we had stumbled into the headquarters in exile of the PLO! The bathroom, if you could call it that—it was really just a small room with an open portal directly into the septic tank—was to our right. As we came back out, still holding our collective breath against the terrible stench, everyone in the place was standing. They started to chant "no bomb Libya, no bomb Libya" as we sidled through the crowd to the door and beat feet back to the cab, thankfully now running again.

When we got across the border and into Annaba, our first order of business was to rent cars. Not as easy as you might think, since they don't have Hertz or Avis. The selection of cars was not only very small, but most rental places (which were mainly just one guy who owned an extra car) wanted to keep a passport, which was obviously out of the question in our case. Fortunately, the Penguini had made up an Arizona driver's license for each of us. They wouldn't have passed muster in the States, not with the little yellow smiley faces, but they worked in Algeria as security for the cars. We finally got one car and left to recon the Algerian side of the crossing point, while Nancy and Jeff stayed to look for a second car. By the time we returned they had procured another one.

The original plan was for Cosmo to head back into Tunisia and man the pickup point at the border, but by now everybody had seen enough of his paranoia that the team insisted Penguini go with him to keep him under control. So they headed back to the border while the rest of us took off on our 400-kilometer journey.

Unbeknownst to the rest of the team, Cosmo and Penguini had a terrible time in customs. For well over an hour the customs inspector delayed them while constantly fingering the contents of their suitcases. Amazingly during this entire time, he failed to find the walkie-talkie that was wrapped up in some clothes in the center of the case. Finally, Penguini placed all his cash on the counter and turned away for a minute. When he turned back, 500 dollars was missing and his documents were stamped. Unfortunately, the whole episode made Cosmo so paranoid that as soon as they got clear of the border, he grabbed the radio out of the suitcase and chucked it into a nearby pond. Penguini was shocked: they had successfully made it through customs, nobody would be looking at them again. This action would have dire consequences later in the mission.

Nancy and Jeff left first, and drove to the rendezvous point, the same beach next to the power station where we were originally going to land. They expected us to be right behind them, but we didn't show up until late that night. We had a flat tire and discovered that tires were a controlled item requiring a government permit to purchase. By the time we found one on the black market, we had lost almost a full day. We spent the next day hunkered down on the beach, while Jeff drove the team through town so they could get a first-hand look at the house and a familiarization with the exfiltration route. That night we went to dinner in Tizi-Ouzou, and at 10p.m. Nancy called the house to see if the children were home. Her son answered the phone and said that his sister was there too. Go time!

We now had three cars, having rented one locally the previous day; one was black, one gray, and one red. We decided to keep the red one stashed out of town along our escape route and drive the other two into the village. At the outskirts of town we stopped and covered the license plates with duct tape, put on our ski masks and black coveralls and headed for the house. We parked both cars along the east wall, close enough to use them as ladders to assist our entry. Leaving Nancy in the backseat of the front car, with her walkie-talkie covered by Arab headdress, the four of us jumped the wall.

The house was actually a walled compound containing two buildings. The main building was two stories tall, and housed the living quarters, the kitchen and the grandparents' bedroom. The second building was one floor with a hallway dividing four rooms, two on each side. We expected to find the children and their father in the second room on the right. We moved down the hall to the far end, where the door to the courtyard was open. We carefully closed it and I was posted as a guard. The other two team members started to work on the locked door with a crowbar, while Jeff nervously paced the hall. The noise they were making with the crowbar sounded like a car wreck in that tight, enclosed space, I couldn't believe that the family was not already up and arming themselves. Jeff felt the same, and his nervousness took him back outside to look over the wall and check on Nancy, and what he saw almost gave him heart failure. The back door of the car was open, and an Arab man was leaning into the car talking to her.

Jeff rushed back to the team to report his discovery; the first question on everyone's lips was "Is it her husband?" Jeff replied that he couldn't tell, but it could be. Then of course the immediate consensus was "We've been set up!" We had a hasty planning session and decided to go out through the courtyard, across the south wall and come around the corner and surprise him. Then get in the cars and make tracks fast, everybody sort of forgot about the kids for a moment. When we opened the door to the courtyard, we saw that the window to the room directly across the hall from the room we were breaking into was open and light was streaming out onto the courtyard. We dispatched Blaine, our tallest member, to check it out and to close the window if he could. He quickly came back and said, "They're in there."

It turned out that the husband and two children had moved across the hall at some point, and all three were asleep with the television on, which probably kept them from hearing the tortured shrieks as we pried wood away from the other door. We decided then and there to go for it. Gary and Blaine would enter the room through the window and subdue the father and take the kids while Jeff and I would go out into the street and handle whatever threat was out there.

We returned to the wall and looked over, only to discover that there were now three men in the street. We stealthily climbed the wall and leaped over onto the hood of the car. In the process Jeff discharged the Taser he was carrying into his right leg by accident and the leg went completely dead. I started off toward the three Arabs, Jeff following, limping along on his dead leg like Chester in an old *Gunsmoke* episode and flashing sparks from the shock baton. The reaction from the three Arabs was amazing. All of them let out high-pitched, bloodcurdling screams and ran into the alley next to the house next door. When we looked into the alley, they were clustered in the back, holding brooms and mops up in defensive postures, frozen like some surrealistic tableau. The Beast told me, "These guys don't look like they want to fight; I'll stay here while you go back in and help the other guys."

While we were in the street, the other two team members jumped in through the window and subdued the father. They had flex-cuffs along and they bound him hand and foot and carried the children out. Jeff stayed at the mouth of the alley, threateningly zapping the test circuit on his electronic shock baton, waiting for his right leg to be functional again, until he saw me holding the little boy over the wall. He rushed to grab him and hand him into the car to his mother, and then ran back to the alley to check on the opposition. They were still there, as rigid as statues. When Blaine climbed over with the girl, Jeff ran to that car and started the engine. Blaine and I got in the lead car with me driving and sat there idling, waiting for Gary to get clear for what seemed an eternity. He finally appeared and we all got out of town in a rush.

What we learned later was that Gary had brought a powerful sedative and he was in the room giving the husband an injection, but the needle broke off in his butt before Gary could administer the full dose. It did make the husband groggy and unsteady, slurring his words like a drunk which helped delay pursuit.

Once out on a main road, every bit of damaging evidence started flying out the car windows into the ditch. Radios (all but one), stun guns, flex cuffs, balaclavas, coveralls, we wanted it all gone fast. By the time we got to the third car we were effectively clean of all our tools and recognizable clothing. We transferred Nancy and her two children into the red car with Jeff driving, and took off for the border in a well-spaced convoy. Blaine and Gary rode point, far enough out to recognize a roadblock, turn around and come back and warn us before Jeff, Nancy, and the kids got close so we could find another route. I rode behind them to offer what support I could. Now all we had to do was traverse 400 kilometers of really bad roads with the local police and gendarmerie after us and make an illegal border crossing. Piece of cake!

We had thoroughly reconnoitered the first part of the route, that led north from the village to the coast road, but from there on we would have to depend on maps. In order to stay off the main highways, we had chosen a route through the mountains that would bring us out halfway across the country and far from pursuit (which we hoped would be concentrating on the airport). As we started up into the mountains the weather got extremely cold. We couldn't believe it, since it was now April, but the landscape looked more like Tibet than North Africa. We reached the pass where we expected to cross the highest point and start down the other side, only to discover that the road was closed due to the excessive snowfall. We had to retrace our steps and find another way through.

After 17 hours on roads that felt like they had been paved with washboards, we finally arrived at Tbessa, a small town near the border, in a blizzard. The snow at this point was coming almost parallel to the ground and all the road signs were iced over. We would stop and I would have to get out of the car to scrape the sign at almost every intersection in order to orient ourselves. When we got into town, we had to find a place where Jeff could fix up Nancy and the children's passports. We had made exit visas from Algeria and entry visas for Tunisia before we left the US, but the dates had to be carefully entered by hand in the middle of each stamp. We found a small hotel and Jeff went in alone to rent a room while the rest of us stayed in the cars.

After a lengthy argument with the clerk to keep from giving over his passport, he horribly overpaid for a single room. The room was disgustingly squalid, with green slimy water half filling the bathtub, filthy bedding and a single low wattage bulb hanging from its frayed wiring as the only light source. Jeff got out the stamps and his other equipment and working on a small end table in the bad light, produced fully functional visas in all three passports. With this accomplished, he came back out to the cars and the convoy headed off through the blizzard toward the crossing point.

By the time we reached the border, the snow had stopped falling but it was still bitterly cold (maybe 15 to 20 degrees Fahrenheit). We scouted the crossing area and tried to make radio contact with Cosmo or Penguini on the other side with the one walkie-talkie we had kept just for this purpose. Unknown to us, their radio was lying on the bottom of a small lake. After an hour of trying, we finally gave up. We knew that if we left Nancy and the children over there without a reception committee, they could freeze to death, so we moved out into the desert, pulled the cars down into a wadi, and hunkered down to wait until the following night.

We awoke at first light the following morning, after a few hours' fitful sleep crammed into the cars. Our first order of business that day was to recon the crossing point again in daylight and finalize our plans for that night. Jeff and Gary drove back down to the border, while the rest of us stayed in a wadi deep in the desert. The crossing point looked innocuous enough, a road with a single gate and a little shack with two very bored looking guards. They scouted the other side of the border with binoculars but saw no sign of Cosmo or Penguini.

At this point the fatigue, fear, and stress of the previous days was starting to really affect everyone. During the drive back from the crossing point Gary almost lost it. He started raving about how we were going to get caught, how we needed to just get out of there any way possible and abandon the mission to save ourselves. Jeff told him that our mission was to get Nancy and the kids out, and only after that was accomplished would we think about getting ourselves out, if we could. He looked at Jeff and in a voice filled with amazement said, "You really mean that don't you?"

It was obvious our little band was unraveling, and something had to be done. Jeff got the team together and told them that to facilitate things, he would cross the border that day, and come back through the crossing point at night to pick up Nancy and the kids. We arranged a rendezvous in a large culvert under the road about 200 meters from the border where we would meet at midnight that evening. The team decided that one other member should accompany the Beast, so I went with Jeff and we drove to a legal border control point about 20 miles south. We abandoned our Algerian rental and walked to the control point. It was very tense, but we made it through without incident. We caught a cab to Kasserine, rented a car and waited until dark to head for the border. Here our plan revealed a serious weak point. This was the northern area that had been reconnoitered by the other group. Neither of us had been to this part of the border on the Tunisian side before, and had to navigate from the map. We started out over small, unimproved roads toward what we thought was our side of the crossing point and wound up driving right into a Tunisian military base!

Before we could turn around and get out of there, armed sentries surrounded the car. The sentries asked us politely for identification, and we turned over our US passports. The Beast tried to explain that we were lost and needed directions to Kasserine. The military held us in their offices for about four hours, offering us coffee and Cokes, and finally escorting us back to Kasserine, escorted by an armored car full of troops. We had no choice but to follow, having already missed our scheduled rendezvous anyway. The soldiers led us right into the parking lot of the Kasserine Hotel, and then parked their APC at an intersection about a block away. We knew that if we were caught back out near the border that night, we would have no explanation, so we decided to just sleep in the car and go back at first light.

Unbeknownst to us, at 5a.m. the next day Cosmo and Penguini left their rooms in that very same hotel, walked right by our car in the parking lot and headed for the crossing point.

At 7a.m. we woke up and headed the same way. In the daylight it was easier to navigate, and we found the crossing point with no problem. We checked to see if the prearranged sign had been left to indicate a crossing had been made the previous night, but there was nothing there. We drove slowly through a nearby village, but all we saw was a single local man standing outside his small house who smiled and gave us a friendly wave. Then we started to panic.

We headed back to the city to find a phone, in order to leave a message for the border team that we were across and needed to link up with them. When I called the company answering service in California (this was before the days of cell phones) there was a message from the team still inside Algeria that "three packages were delivered last night on schedule." We were in our car and broke land speed records getting back to the border. In a panic we searched the area on our side, finding nothing. We were completely mystified as to what had happened; they obviously were not here.

I started to fear that in their haste to get out, the other team members had left them at the rendezvous site on their side. If that was the case, they were now sitting in a culvert under the road 200 meters inside Algeria. We decided that we had to go in as soon as it got dark and check. Meanwhile we returned to the Kasserine Hotel and ran into Cosmo and Penguini. They had also heard the message, but they had been at the border at 5a.m. and spent an hour searching. We told him that we had been there at 7a.m. and again at 9a.m. and had likewise seen no trace.

We sat over coffee and told them about our plan to go back into Algeria that night. They told us that it was impossible to approach the border at night, so we would have to be dropped off during the day and hide until nightfall. We decided to go back at 3p.m. Just before we were ready to make one more visit to the crossing point, Cosmo said that we should call the answering service one last time just to be sure. We placed the call and much to our surprise, the message was that Nancy and the kids were in our original hotel in Tunis. We raced the 150 miles north in record time. When we got there and found them snug in a room in a four-star hotel, we all felt like a thousand-pound weight had been lifted off our chests. We still were not out of the woods though; our other two team members had not yet arrived, and we still had to get out of the country.

We asked Nancy exactly what had happened the previous night. She told us that when we didn't show at the rendezvous point, Blaine and Gary had brought her and the kids across and left them in a field behind the village, but they had not taken the extra time to load the signal. At 5a.m. when the first team arrived, they were still asleep in their sleeping bags out in the tall grass. By 7a.m. when we had arrived, they were in a house in the village being served breakfast. This was the house of the man who had waved at us as we drove through. By 10a.m. when we went back the second time, they were already headed north in that gentleman's son's pickup truck.

Later that evening our other two team members showed up in bad shape. Gary was almost incoherent and had actually developed an uncontrollable facial tic. After they had brought Nancy and the kids across, they had headed to the same border point Jeff and I had used the day before. They cleared customs and immigration without incident on the Algerian side and were in process on the Tunisian side when four carloads of Algerian police had shown up demanding their immediate return. The authorities were told that they were wanted for questioning involving a kidnapping, and for four hours they sat in a small concrete cell while officials from the two countries argued their fate. Finally, the chief of police from Kasserine was called, and told them, "They want you returned, but they have no real evidence. I will let you stay in my country, but I suggest you do not stay long." As soon as they were released, they made straight to Tunis and wanted out of the country like right now! Immediately!

We knew that they were stressed almost past breaking point, so Cosmo took them to the airport and got tickets on the first plane out, which happened to be a Lufthansa flight going to Frankfurt, Germany. He told them to make their way to London and we would meet them there tomorrow. Penguini also left, but he didn't want to be on the same plane, so he caught a Tunis Air flight to Barcelona, Spain. Jeff, Cosmo and I stayed with Nancy and the children. We had a scheduled flight to London early in the morning.

The next morning, we went to the airport with a prearranged plan to break up into small groups and check in for the flight separately. Since Nancy and her kids had Jeff's homemade entrance visas in their passports, he was a little worried about them going through passport control. We told Nancy to go first; the rest of us would hang back and watch so we would be in a position to offer whatever help we could if something went wrong. The departure lounge was up an escalator, so we watched Nancy go up and about five minutes later Jeff followed. The passport control and customs area was empty, so he went on through into the departure lounge and breathed a huge sigh of relief, prematurely. Unfortunately, after searching the entire departure area there was no sign of Nancy and the kids. Jeff was incredulous: where could they have gone? His only thought was that they had been picked up at passport control and were in the small police station which adjoined it, being questioned. His heart sank to think that we had made it this far only to lose now. A few minutes later Cosmo came into the lounge, all smiles and giving Jeff a thumbs-up. He walked over and said, "Well, we made it." Jeff told him, "Yeah, we made it, but Nancy and the kids didn't." At this Cosmo turned white as a sheet. "What do you mean?" Jeff told him that he had searched the entire departure area and they were nowhere to be found; his only explanation was that they were being held in the police station.

At this point Cosmo decided that he could be much more useful working through the US Embassy in London. "Why don't I fly out and contact the American authorities and work that way, my talents would be much more useful there, and there is really nothing I can do here." The Beast told him to forget that, the only way anyone was leaving was when we all did. Then he told him that since he was not in Algeria where the "kidnapping" had taken place, and had the stamps in his passport to prove it, it was only logical for him to walk into the police station to find out what was going on. Walking in there was probably the most difficult thing he had ever done, but he sucked it up and headed in.

Amazingly, when he got inside, they weren't there either. About 10 minutes later Nancy and her kids came out of a snack bar across the hall (which Jeff and Cosmo had never even known was there) and passed through without incident. I was still trailing them; I had seen them in the snack bar and kept an eye on them the entire time. I came through a few minutes later and we all boarded the motor coach that would take us to the plane. As we stood on the coach, thinking that we were as good as home free, Nancy was paged on the PA system and panic clutched at our bowels once again. Cosmo took his French language skills forward to ask the driver what it was about, but it turned out she had left her tickets at the check-in counter. More adrenalin to process.

Unknown to us at the time, the Penguini was having his own issues. Penguini had booked himself all the way to London on Tunis Air, but he had a change of planes in Barcelona. He had to get his bag and go through Spanish customs and then go check-in again at the Tunis Air counter. When he got there he noticed two suits who seemed to be looking through passenger manifests and shuffling papers behind the counter, so naturally the paranoia gene kicked in and he was absolutely sure they were looking for him.

Penguini ran over to British Airways and blurted out the whole story of our adventure in Algeria and his suspicion that the authorities were looking for him in a 30-second, rapid-fire stream of consciousness. The two matronly British women behind the counter just looked at him slack-jawed until one asked, "Are you serious mate?" "I'm absolutely serious," he replied, "I need a ticket on the next flight you have to London." She responded that they had a flight about to leave and she could get him a ticket immediately. "Do you have pesetas? We only accept Spanish currency."

Penguini only had US and Tunisian money so he had to sprint the length of the airport to a Cambio. When he got there the British woman had called ahead and the man in the booth gestured him to the front of the line. He threw down a handful of hundred-dollar bills and scooped up a handful of pesetas and ran back to British Airways where the ticket was waiting. He went to grab his suitcase, and it was gone. He started having a meltdown.

The counter clerk calmed him and pointed up the stairs that led to departures where a British Airways steward was carrying his suitcase and motioning for him to hurry. This of course is one of Britain's exports to the rest of us, the ability to have aplomb in the face of insanity. She had made a few calls, pulled the right strings, and zip, Bob's your uncle, his escape route was clear. He bounded up the stairs and rushed to the gate, only to find the jetway had already been retracted. The steward threw the suitcase across into the plane, looked at Penguini and asked "Are you up for it mate?" With that he backed up, got a running start and jumped across the open space into the plane.

Now the steward was a tall, lanky young man, while Penguini was considerably older and with much more compact dimensions. He backed as far up the jetway as he could, took a long run and launched himself across the void so violently that he not only cleared the space, he collided with the opposite bulkhead hard enough to knock himself out. When he came to he was belted into a first-class seat, airborne on the way to London.

We finally made it on board; Cosmo and Jeff were in the last row of the first-class section on the left, I was across the aisle from them. Nancy and the children were in the first row of the second section, immediately behind Jeff and Cosmo and separated only by a hanging curtain. As we sat there waiting for takeoff, Jeff noticed an official-looking pickup truck screaming across the tarmac and coming to a screeching stop at the boarding ladder. Two large gentlemen wearing black suits and aviator sunglasses came bounding out and ran up onto the plane. While we attempted to shrink down into our seats, one of them had an animated conversation with the crew as the other carefully scanned the passengers. It may have been routine and had nothing to do with us at all, but once again everyone's adrenal glands went into overdrive.

After about five minutes the two men left, the doors were closed, and the aircraft taxied out onto the runway. Never has any group of people been so happy to feel landing gear go up. Once we were airborne it finally hit us how close we had been to disaster, but we had pulled it off. We have been on other missions of this type both before and since—some were easy, some difficult, but this one was the most intense, most confusing and ultimately the most satisfying of them all.

* * *

Jeff winds up giving an interview to the local paper, which goes national. That's probably how the people in Washington tracked him down for the phone call. He figures it will lead to more operations like this one and it does; maybe we'll cover a couple of them in a future book. Penguini then goes and gives a long fanciful version of events to some reporter up in Nevada somewhere. He explains the many embellishments as being due to the fact that the reporter was really good looking.

CHAPTER TEN

Pancho and Lefty

Not long after our return, I receive a call from my good friend Francisco in Mexico. He is a tall, handsome Lebanese Mexican with courtly manners. Francisco and I had met when we had been training bodyguards for some of the wealthiest families in Mexico. He had an armoring plant and his firm was one of the premier manufacturers of armored cars in the country. I needed vehicles for my clients and found that the US companies would sell you a car, but their customer service sucked. I had settled on Francisco's company because it was quality and they went the extra mile to ensure that everything was right.

His call relates to a request that he has received from an American, a former security guard at the US Embassy, who has set up a private security consulting company. I have met Ricardo several times and found him to be competent and professional. Where my friend Francisco is urbane and sophisticated in manner and dress, Ricardo is the exact opposite. He dresses well but comes across as a mix between a carnival barker and a barroom brawler. When you wear all the hats in a company you have to have multiple personalities. He has a busload, but he is good at what he does and more importantly, he is honest and trustworthy.

He has a client that needs help, since his son had been kidnapped on his way to the university a few days before. The kid did everything he had been told not to do. He was in his car, with the window down and the door unlocked, stopped at the intersection of the Barranca del Muerta. There were three routes to the university that Ricardo had outlined for him to use at random so not to set a pattern. The kid had ignored that and used the same one for several days.

He was waiting at the stoplight with a traffic cop not 50 meters away when two vehicles boxed him in, two men ran out and pistol whipped him to the floor, then one drove while the other subdued the victim in the back, pushing him to the floor. The police officer's reaction had been to turn his back and wait for the gang and their victim to speed off. This was a classic technique and executed with military precision. They were gone in 30 seconds with their victim. Most people who had witnessed it wouldn't tag it as a kidnapping, it happened so fast.

The case had proceeded in the usual manner, the family was called and told the ransom demand a day after their son missed coming home, and was told that the kidnappers would be in touch. The family's attorney had already called Ricardo the minute they discovered him missing.

Ricardo and I had been discussing the aspects of kidnapping in Mexico and how it was becoming a business. At that time, the average ransom was around 100,000 dollars to one million dollars, depending on the victim's financial status. We had been tracking trends and the average length of a kidnapping was between 21 and 35 days, start to finish. The usual method was pictures of the loved one or a phone call with the victim pleading for help to the family, followed by instructions and a warning that if the police were involved the victim was toast.

The gangs were fairly confident on uncovering any police involvement as 95 percent of the kidnappings featured police involvement with the gangs, from the common street cop to high-ranking commanders. Law enforcement had been purchased at all levels. This case has all the earmarks of a sophisticated and well-disciplined group. We are about to find out just how well versed and disciplined they are. As Ricardo is briefing me on the kidnapping, and the subsequent actions the family have taken, some bright spots emerge. They had gotten the case moved to a special investigation unit within the State of Mexico's State Police. This outfit had been set up specifically to combat this type of crime and its members are hand-picked and vetted every month.

The commander is a former surgeon, with plenipotentiary powers so that he can investigate and arrest any government or law enforcement official connected to a case. Ricardo and I talk about using some of the state-of-the-art technologies to outdistance the learning curve exhibited by the gangs. The gangs have their own tech arms that are familiar with tracking devices and dyes, so their drops are often equipped with Faraday bags to put the money in and prevent the tracking devices from signaling outside. This can be a metal container or metallic cloth. But we had recently discovered a new method which was a chemical taggant, with a sophisticated detection model.

Ricardo wants to use it on this case, so I fly down and get introduced to the family and to the State Police commander. He had done a thorough job in vetting both the police and all who could be exposed to our activities. The Mexican criminal code states that if you give money to the kidnappers without having the police involved you can be charged as an accomplice, but now Ricardo has eliminated that by locating this unit. Most families know the police are involved in some way with the crime, so they are reluctant to let them inside their homes or into their private affairs.

The police unit is superb. I find them to be dedicated hard core cops, vastly underpaid, but motivated to the highest degree. Their track record is unsurpassed. But they often see their arrests corrupted by the courts with the convicted getting off with light sentences and/or recruiting new talent from the prisons while they were there.

The family wants us in and that is allowable by both Mexican law and their kidnap insurance. Most people don't understand how kidnap insurance actually works. The insurance company doesn't give you the money to pay the ransom. You must raise it or borrow it yourself. Then the insurance company pays you back, provided you follow their rules. If you launch a private rescue without the police being involved, if the victim dies because of your actions they won't pay. They will however pay for outside consultants, as long as they are working in concert with the police. That is the category that we fit perfectly, so the insurance would pay the family back both the ransom and the fee for professional consultants or other providers.

Ricardo starts by having me get familiar with the police unit while I am there. I go to the range with them during one of their rehearsals and like bad dogs everywhere they put on their best show. I am invited to try my hand and soon convince them that not only is my resume correct but that I have done this before. We bond well. The exercise is the human version of sniffing each other's ass, as dogs do.

I fly back to organize the team and equipment that I will need for the mission. My first contact is with the Jewish version of Q in the Bond thrillers. Brilliant, frenetic, and extremely competent, he is the perfect source for what we have in mind. He had approached me months before about being able to set up a reliable test of the concept, but I didn't have an applicable situation at that time. Now the family and the authorities are in position to act and act in a timely manner, so I am confident that we can provide the service.

My first step is, with him, to arrange for a technical team to accompany the detector to Mexico, followed by delivery of the taggant. I also need assistance from someone I can rely on to shepherd logistical aspects. I opt for the Beast because I can lay it all off on him and concentrate on the mission. As the mission progresses, his role will expand.

When I return to Mexico the situation has progressed. Originally, they had a trained negotiator who was acting as the go-between for the family. They have gotten rid of him and the mother of the victim has taken over that role. She is superb at the job, to the point the gang is calling her Iron Britches. She has maneuvered them so that she can guarantee that her son is safe and that he will be returned whole. This has also given the police a chance to locate the area from which the gang operates. By their modus operandi they have identified the group that they believe has the victim. The bad news is that the gang is one that is all too well known to them. It is led by a former Special Forces officer, and they have been tracking him for two years trying to kill him. He is smart, well trained and ruthless, he enforces iron hard discipline on his minions, and the penalty for failure is execution.

We have arranged to do the drop as described. The police are set to take up surveillance on the drop point, which will be off one of the major toll roads going out of the city. Classic drop point. The gang will surveil the road and the dropoff point which will be just off the road. They will be looking for any vehicles that don't fit the pattern and will kill the victim if they suspect any police involvement. The gang is working out of Texcoco, so the police know where at least one of the lieutenants lives and a couple of their safe houses. They have opted for a very loose surveillance on the drop point, with a screen which we will be part of, along with the detector. When the vehicle passes us, we will notify the police then follow the vehicle at a distance.

The manufacturer of the detection device has sent two operators with the machine. Both are former US Marines. One is a short, compact individual who is very competent and follows instructions to a tee. The other is a former staff sergeant who is full of his own importance, disdainful of the Mexicans, and voices his contempt and distrust frequently, wrongly assuming that none of them can understand English. He is also a klutz with the equipment, so I keep the younger of the two operating the machine, and the other outside pacing around. He is starting to be more than annoying but the mission is in motion so I have to take him along.

The drop is made as we wait at the chokepoint and after about two hours the taggant detector lights up. We follow the scent to the outskirts of the city to where the victim is released. First phase success, and that is the key for us to close in on the gang.

We return to our hotel, which is across from the US Embassy, and we are enjoying breakfast when the *Comandante* and his two lieutenants arrive and say that they are ready to make the raid on the gang's location.

Right there in the dining room the *Comandante* hands me a ballistic vest, telling me to put it on as we get up to leave. I am struggling to put the vest on and he hands me a pistol for, as he puts it, "self defense." The other patrons of the restaurant are staring at us as if we are something not included in their travel brochure. The head waiter is reassuring all and sundry that we are Policia, not bad man.

I have had the Beast tracking events and staying out of sight to prepare a back door in case things go awry and he has been coordinating with his "friends" at the embassy and with our own private sources. He has managed to do all of the above and more, so that I know that we have loose cover on us as I leave the hotel with the SWAT team. He gives me a drop of the paper he has been reading in the lobby, so I know that he will shadow us to where we are going and make the necessary arrangements for the SOS call if we need it. He fits the role perfectly. He affects the bored disinterested background enough that you would dismiss him if you were scanning the area. Even the police miss him. More importantly the problem child that came with the machine misses him completely. This is not surprising since he insists on staying with the machine which is installed in a waiting van. That's fine with me because I have gone beyond the point of annoyance at his attitude. The other kid, who I have nicknamed "the Combat Midget," takes it all in stride.

I get into the *Comandante*'s SUV with two of his gorillas, one of whom is nicknamed "Mouse," and I get a chance to look carefully at the equipment they have given me. I look at the vest and it is old enough that I am sure that it probably won't stop all of what it was originally designed for, but it will do. The feeling of protection is similar to what you feel wearing a cup in a rugby match. It's there but if the 800-pound gorilla runs you over, it's mostly illusionary. Next, I notice that the serial numbers on the 9mm have been ground off. It's a good weapon but this did not come out of the armory, it probably came out the evidence room, so it is a throwaway. I catch the *Comandante* looking at me and he gives a knowing smile and a shrug. I don't care; it's an honor to be asked to participate.

The techies are in the van with the equipment. We convoy out to Texcoco on the toll road. We pull off and I transfer over to the van to oversee the detection operation and stay in touch with the police. We are directed to follow the lead vehicle and they will give us instructions en route. Several more SUVs with additional officers from the unit join us. As we pull out, I see another vehicle sitting at a rest stop and recognize the Beast sitting in the driver's seat, eating a burrito from a road-side stand. We flash by and follow the road to what appears to be a cluster of farm buildings off the road to the left. We have been getting trace amounts on the detector. Then, when we are abreast of the buildings, we get a spike. We don't stop, but four of the SUVs do. The Beast almost gets caught but turns down a side road and doesn't attract attention. My cell rings and it is him. We turn into the city and I tell him what road we're on and he catches us five minutes later and falls back into the traffic, into the drag position. The vehicle he has is perfect for blending in. It's a late model with the appropriate dents and dings, nice neutral color with a layer of dust. I'm wondering if he rented it or hot-wired it.

When we get to Texcoco the *Comandante* asks us to follow him past the police station, prior to going to the gang's hideout. When we pass the police station, we get another heavy spike in the readings, indicating either the money is in there or people who have handled the money are. He notes this and has us go by the addresses that they have determined to be sites where the gang congregates or where leaders live. We note each one that gives off readings, and relay that back to him. At each point they drop off a surveillance unit to eyeball the place. I am watching their procedures and tactics. They are not using the local police for a reason.

We continue out of town to a truck park and stop to reorganize for the raids. The *Comandante* has four units and four targets. The one outside of town will be taken down at the same time that we are hitting the main target where they believe the gang leadership is. The police station and one other will be raided by state police and the special units. I transfer over to the *Comandante*'s vehicle and everyone starts putting on their kit. I will be along as an observer right behind the assault team.

Like all operations there is always a last-minute glitch. Ours is in the form of a news team that is doing a story about crime and is taping man-on-the-street interviews about two blocks from the targeted building. They are about to get the story of a lifetime dropped right on top of them. The police unit has two of their admin people ready to interface with the TV crew to make sure they don't get in the way when the shooting starts. I spy the Beast down the street blending in with the foot traffic.

We roll up to the house and the assault element makes entry on the first floor and simultaneously on the second floor via an outside staircase. At the same time more state police roll up and start cordoning off the neighborhood. There is immediate gunfire from in front of me inside the building and I hear shots from downstairs. Lots of shots, a stun grenade goes off then more shots as the *Comandante* and I enter the room. There is still gun smoke and smoke from the flash bang filling the room, making it hard to see. They are throwing open windows and checking the inert figures on the floor. There are more gunshots from a back room then the sound of breaking glass. Two team members come down the hall from where the shots came from and tell the leader something I cannot hear. He shouts out orders and the team begins to lock down the site.

There are four bodies in the center of the room, lying around stacks of the ransom money, which they had been counting. There are two more bodies lying at the foot of a blood-spattered wall. One of them is still a wiggler, shot in the gut and one leg. Just out of arm's reach, on the floor there is what looks like a Walther MPK submachine gun covered in blood, and what looks like a bullet crease across the receiver. This gun was issued to the navy and some police units so it's not out of place, they probably purchased it on the flourishing black market for such hardware. The team begins to flow back, searching anything they may have missed during the assault. It's textbook procedure and they soon have the scene secured. Judging from the radio traffic the building and area around it have been secured. The *Comandante* gives a signal and everyone files out of the room except two with the wounded gang member. No one is bothering to give him first aid, but he is nearly gone. He is starting to froth blood from his mouth and nose. He must have caught one in the lungs.

We emerge from the building by way of the stairs and make our way down to the street, where more police from the state are showing up. I go across the street and sit down on some crates in front of a storefront. I open the vest and try to come down from the adrenalin rush that always accompanies these affairs. I had been on the tail end of the assault with the command element consisting of the commander and his sidekick "Mouse." I had been close enough, but not in the line of fire. The handgun had been in case the situation started to get ugly. At least they have given me a functional weapon (I checked).

I look down the street and see the Beast being questioned by two police officers. The commander walks up, and I tell him, nodding in the direction Jeff is being questioned, to inform him that the Beast is one of mine. He smiles and says I know; we have been following him since we left the first stop. He says something in his radio and the two cops give him back his papers and let him go. He moves off down the road. We will meet up later.

The commander looks at me and says "Poncho and Cisco?" Then chuckles. He notices I have the shakes and tells me that it is the effect of adrenalin, and I look up at him and start laughing. He asks me why and I tell him that today is my 50th birthday. I add, "I've got a bulletproof vest on, a handgun with no serial numbers, we just got done doing an armed takedown: Whoooeee. They don't even make oral sex this exciting where I come from." He just shakes his head and tells me to come with him, that we have more work to do.

There is a crowd of onlookers along with the news crew towards the back of the building where they are loading two figures onto gurneys from an ambulance that had obviously pre-staged. One has an obvious compound fracture of the tibia, and is screaming, the other is out cold with a broken arm and an ankle pointing the wrong direction. These two had been in the back room on the second floor and decided that they would take their chances through the window to the street below. They are loaded into the ambulance and the commander and I get in his vehicle and the assault element falls in behind us. We go to the police station where the entire shift is gathered outside and disarmed. He asks us to scan the building again then begin to bring out items and put them one at a time through the process. They are articles of clothing, personal effects etc. All of them are hot.

We spend another two hours and then the *Comandante* tells one of his team to escort us back to the hotel. When I get there, I check to see if one of our number is no longer in Mexico. Prior to the raid we had an unfortunate incident. When we were staging, the two Marines had taken our vehicle to scan a house and while they were standing outside, the police with them had to run a radio a few short blocks to replace one that had gone down. There was no time to explain, and the police had simply driven off with the vehicle. It came back some 15 minutes later. During that time the older of the two had started to rant and rave that the Mexicans were stealing the technology, compromising the machine yada, yada, yada; all within the hearing of the entire team. I had worked with these guys long enough to know that they all were able to understand enough English to know that what he was saying was not just rude but insulting. They had given him the skunk eye and the commander, who spoke fluent English, had also glowered at him. I had defused the situation by going over to the moron, who was still spouting his bullshit, and told him to shut up. He looked at me in shock, but I think he could read that was the end of that act. I hissed at him that the team understood him perfectly and if he wanted to make it back to the US to shut up. I called up Ricardo, explained the situation to him and told him when he got his hands on this moron to put him on the first flight out of Mexico, no matter what the destination. I told the ranter to go back to the hotel, pack his bags and wait for my partner to meet up with him. End of incident, he was gone by that afternoon. A stupid move by a stupid man. The other techy performed his duties to a tee.

We spend the next few days winding down our involvement with the case. Jeff takes a flight back a day before I do. When I got back to the hotel we had a short brief back, then I had gone on to see Ricardo and the *Comandante*.

Ricardo, my partner in this whole affair who had arranged our contract and payments, was there to help tie up the loose ends. He reminded me of Don Rickles' character in *Kelly's Heroes*, Crap Game. Add a couple of inches in height, he had the same mannerisms as the character that portrayed the sergeant that ran the supply depot in the rear, making a sideline of the black market in cigarettes, nylons, booze and the occasional object d'art or bullion.

He is Crap Game in that regard, leave no stone unturned, no change on the table. A very capable individual who was invaluable in both dealing with the family and interfacing with the vetted unit we used to process the case. He had been in the US Army before he became an embassy guard and had parlayed that into a successful business as a consultant. He lives in Mexico City, is married into the culture, and was able to get us what we needed when we needed it. Between him, my partner in the armoring business and our shared *padrino* with the law firm, and the Beast being on the edge, I had been sure that we had a back door if things went sour.

The wiggler never made it out of the room, they just let him bleed out. The other two would be imprisoned for the term at that time of five years. They would eventually be released, and both die in a later kidnapping. The ex-army officer that ran the gang continued on for a few years until the *Comandante* and his crew trapped him and killed him in the shootout.

One of the bodies on the floor had been the assistant state attorney, who had been counting out the cash when the whirlwind came in the door. The precinct of the police department was all arrested since every one of their lockers showed the money had been distributed in part to them. The only two officers not arrested had been on vacation.

The assault team I hold in the highest regard, they put their lives on the line and in many cases paid for their own gas and cartridges. They were taking down five gangs a week and still had only scratched the surface. I found out a few years later that "Mouse" and another of these brave men had been killed in the line of duty.

A last footnote is that after the gang grabbed the kid, they gave two lower-ranking members the duty of disposing of the kid's car. They had instead deviated to do some robberies and got caught by the highway patrol in another state. Once the highway patrol had the car, they had processed it. Two of their number then tried to blackmail the family by saying that they found pornographic evidence of the son engaged in homosexual acts and demanded money for destroying the evidence. They went to jail for their deceit. The two gang members who disobeyed orders were soon found beaten to a pulp and then murdered. No doubt it was the *Jefe* exerting control over his *sicarios*.

CHAPTER ELEVEN

By the Dawn's Early Light

The concept that we float on a sea of coincidences, linked only by our presence and some omnipotent power's whim, has always intrigued me. I must admit that I have personally made some catastrophic bad decisions in my life, and still managed to come out the other side at least with my dignity. The same can be said for my comrades. It seems that something watches over us poor fools, letting us dangle like a windsock, until the wind of fate blows us on our way.

We had returned from Mexico, having witnessed first-hand what happens when the rule of law breaks down. Mexico with all its charm and glory is fading whilst the barbarians sack its soul. However, we are satisfied we had earned our fee, and settle in to finding more work. I had given the inventors of the detection method a brief on how the binary system performed. It was a useful tool but needed some tweaking to make it more user friendly. We had also established a relationship with people that could vet whomever we were working with to insure we didn't get a penetration from the cartels. We had a track record so perhaps there was more work in that direction.

I had picked up some work from a mining company, or as they prefer to be called, "resource developers," to do some security and investigative work. The first is mostly inspection, assessment, and directing the change; the second is more complex depending on what the target is. In both cases collaboration with one's playmates is essential. We are a varied lot. We have been together so long that we are like some twisted Arkansas hill tribe. We also have our own networks from times past that we acquired during our travels together and solo. When it comes to a question of depth of knowledge, and Jesuit logic, the Beast is in his element. If you want to survive and prosper in hell, you need the Penguini. Together we will take on just about anything especially if it pays. We all have done charity gigs before, but eating regular is much more appealing.

I had spoken to Jeff a few days before Halloween, and the Mexican equivalent the Night of the Dead, remarking about how the crazies seem to proliferate in the fall. He of course has some Freudian or Jung-based theory mixed in with

Aztec poetry, that explains the reason for chaos. Me? I am just a dimwit, but a lucky dimwit.

There is always a lag in our business with the government, as they start drinking sometime before Thanksgiving and don't come out of their booze-and-turkey coma until January fifth. So we tend to forget about even trying to look for work in that sector during that time period. But fate seems to explode on the scene from the private sector during this lag, and it is never a pleasant surprise.

I receive a call from Mexico from my friend Francisco. He is so courtly; I always have the impression of speaking to a Spanish grandee. His impeccable manners and laid-back demeanor are a breath of fresh air. This time, however, he is obviously agitated and concerned. He gives me the details of his unrest and asks if we can help his friend who has a dilemma.

His friend is a customer who buys armored vehicles from him and is the owner of one of Mexico's large agro-business firms. His father had been kidnapped a few days before and they need help in recovering the victim, working with the police. I listen to Francisco and tell him that I will get back to him. I speak with the man and advise him that we might be able to provide technical assistance. I also make the stipulation that we have to have the government's assistance and permission, and that is difficult since there is so much graft and corruption that a leak is inevitable. He assures me that he has the governor's complete cooperation and that the police unit will be a state unit, not the locals.

The stipulation is necessary because every kidnapping has local police involvement, so keeping a secret is going to be a chore. I talk to Jeff about helping me with the operation. He is tied up with work but could help do some of the background until he can break free. I want him as our pivot point in Mexico City so I can get a timely response to my pitiful pleas for rescue if it goes full Penguini on me. The Penguini himself is off on one of his disappearing acts where he communes with his past and present demons. I imagine him wandering the land, a cross between a Shao Lin master and a Klingon, making someone's enlightenment a painful journey. Most likely he is running for his life being pursued by some ursine beastie that raided his meditation camp in the wilderness. Either way I need to gather up the technology and an operational team and I have to do it fast.

Jeff will do the background work and help with the planning. Short of firing up the bat signal and shooting it at the cloud base over the wilderness, the Penguini is out. I need to replace that talent and I have a good idea of whom to call. The first is an old friend from the war and the bank note recovery—Reek. He has the necessary snap switch from idling to full-on massacre if needed, and he is Mensa intelligent. The second choice is a rising talent in our profession. We had trained him during our contract with the IACP and he has grown into a star in the Richmond PD. He has two valor awards and is capable of dropping the hammer on someone in a gunfight. Smart as a whip, and adaptive, I know he could help handle the technical aspects.

This time I am not going to use the technicians as before; we will train ourselves to operate the equipment. This is the same theory that chimps can fashion tools, but I am reasonably sure we can get the technique to work.

I call both and made a compact, secure a flight for Reek to Mexico City and set out to pick up Mick the cop, then continue on to pick up the machine I have purchased along with a crash course in its operation. The client has gotten the loan of the governor's private plane, a Lear complete with crew. It flies in to pick me up and I continue on to Richmond to pick up Mick, landing there around 6p.m. The governor's plane has a huge emblem of a red lion on its tail, the symbol of the state.

As we taxi up, I spy Mick with his new wife, standing by the fence near the gate. He has his bags packed because we are taking off immediately for New Jersey where the equipment is. His wife, who considers me a bad influence and in the same category as foot fungus, turns to him and upon seeing the red lion emblem, snorts and tells him that she would bet good money that it is a magnetic sign. We soon are on the way to the manufacturer, where we spend the next day going through proper operation and troubleshooting the equipment. Then we are on our way to Mexico. I like riding in an executive aircraft, it takes away all the stress of being crammed in a metal tube with 200 other members of the flying public and their various afflictions and bad attitudes. I don't mind serving my own drinks either and the booze is of better quality.

We are united with Reek who is his normal pleasant self, bordering between a Jurassic raptor and a gut-shot rabid wolf. We move the operation into the compound of the client and use his guest house as our habitat. Next to the guest house is a huge garage that holds his car collection which is mostly race cars that he has driven in competition. There are lots of Porsches and a few antiques. This is where we will install the equipment in a van provided by the client.

Next to the garage is a caged area, with 4x4 steel square stock as its walls, embedded into the concrete floor. Crouched down viewing us like menu items is a full-grown lioness. Sure enough, her name is Elsa. She looks sleek and well fed. There is an old Mexican guy that tends the guest house, garage and Elsa's playpen. We ask him why the lion is there, and he tells us that it's much better in there, than loose like when it was growing up. I guess when it reached its maturity, they had some unfortunate experiences with the lion trying to drag the slower members of the staff off into the bush. Henceforth the lion only gets out when the client wants to walk the cat around the grounds, which are extensive. That's fine with me—I have enough excitement without adding misplaced feline rough play to my dance card—but Mick is fascinated with the lion. I keep telling him to leave the friggin' cat alone, but he keeps going out and tossing it food scraps in the belief that it will become his buddy and they can frolic in the garden together naked or something. I repeatedly tell him that all cats are treacherous, and if he doesn't believe me, to consider his relationship with the fairer sex. My warnings fall on deaf ears. Reek just looks at him and says, "That is a wild animal and you, my friend, are merely a 250-pound Osso Bucco."

A fourth member of our team arrives, who is a recently discharged member of Special Forces. He is a younger version of ourselves, perhaps a bit less jaded but willing to try out the adrenalin ride. We also get introduced to the police unit that will be handling the recovery. To a man they are carbon copies of the group we worked with in Texcoco. The state police organs have all formed special units to handle the kidnapping epidemic. They all have to take a polygraph once a month to insure they haven't been co-opted by the bad guys and all are bone-hard lawmen. We bond easily because it takes the same type A personality to excel.

They have been in contact with the kidnappers, and are stalling until we are able to set up and be operational. The payment and the drop are set up for the next day, so we have a limited time to mark the money and to fine-tune our operation. And time to get some rest before we will launch. Once it starts, we will be in that van until the gang is run to ground. The police won't move until the victim is released so we are 24 to 48 hours from when we drop the money off to starting to close in on the gang.

We are going to use the newest member of our team to deliver the ransom money. The kidnappers have given precise instructions for the delivery. One man dressed in white pants and shirt will drive an equally white pickup truck with the bed removed so they can be assured there is no one but the driver in the vehicle. They instruct him to drive to a specific location and retrieve a cell phone they have secreted, then follow the instructions given when the phone rings and he picks up.

We are fine tuning the van and apparatus in the garage. I am securing a sensor head to the roof rack and Reek is doing something inside. Mick had been there just moments before, and we had given him a hard time for the clogs that he was wearing, pointing out that sissies and members of the Euro-Trash set were the only ones that wore them. This had sent him off muttering about our being cretins. We hadn't paid any attention to him after that, focusing on our tasks instead.

I sense a faint sound, just out of hearing range, so I stop and listen. Reek has heard it also. We strain to hear, and we can make out faintly someone calling "Help." We stop and follow the sound, the cries for assistance becoming louder. We finally walk out of the garage. The sight that greets us is both hysterically funny and about to be tragic at the same time. In front of us and lying on the ground on his back is Mick, with his right leg extended and the foot inside the cage. Miss Kitty has bitten through the toe of the shoe, just missing his big toe and trapping his foot inside. She is tensed up with her muscles rippling in excitement, and every once in a while gives a little jerk, trying to pull Mick through the four-inch gap between the bars. Mick has his left foot braced against the wall next to the first bar and is straining with all his might to keep from being a large kibble snack. He is sweating profusely as he looks up at us imploringly. Reek looks down at him and asks him where's his camera that he has been using to photo grab everything we were doing like some Japanese tourist. It is in his photographer vest, the highlight of his Wild Kingdom ensemble. Reek makes him take it out and he takes several pictures of Mick with his foot in the lion's mouth.

The comedy is waning as Elsa is losing patience with playing with her meal. The old Mexican who cares for the cat calmly walks up, barks something at the lion, and swats her on the nose with a roll of lottery forms. The cat spits out the shoe, releasing Mick who collapses where he lays. The cat coughs and makes a face then sulks over to the back of the cage and looks back royally at the four of us. Despite our repeated warnings Serpico had continued to come out here with little snackies for his lady friend and had been teasing her by sticking the toe end of his shoe through the gap in the bars. He had faith that he could slip his foot out of the open-backed clog. He hadn't counted on the speed in which a lion can move. She had trapped his foot quicker than he could soil his britches, then had begun playing with her new toy.

Eventually she would have tired of that and just dragged his body through the bars into a nice paté on the other side. He is grateful to be alive but still pissed at us for the pictures and laughing at his predicament. The old Mexican just shakes his head and continues to rebuke the lion, muttering about stupid gringos at the same time. Elsa merely looks at Reek and me as if to say, "You want some of this sucker?" Then rolls over on her back, never taking her eyes off us. I am thinking how perfect it would be if we could give her a mild sedative, load her in the bag as if she were the ransom money, then drop her off for the kidnappers to have "that magic moment" when they open it.

* * *

It is getting near dusk and the operation is about to begin. We transfer the money into the stripped-down pickup and give our guy a burn phone so he can have it on his lap and the police can listen in on the device the kidnappers will give him. He doesn't look all that chipper but girds his loins and mounts up. He puts on a brave face, but I can see that I am not his favorite person at this moment despite assurances that the police will have surveillance on him the whole time. The police have air assets on hand; I am hoping that they have a handle on the over watch.

We mount up and follow the police in the van with the detector. We are about 10 minutes behind the truck with the ransom money. The radio reports that he has picked up the cell phone and they are following him in discreet formation alternating between the air asset and the ground force. We wind our way through the city in what is an obvious counter surveillance route where the gang can watch traffic behind the truck and report in. There are multiple left turns then right turns in sequence, all designed to pick up on any vehicular surveillance. The commander calls and explains that they know who this gang is and that they have used the same route and technique before so they had a good idea where the drop point will be. They are setting up an area surveillance and will tell us when to come in and start tracking.

We soon move out into the countryside, and we are listening to the surveillance units giving their reports to the command element. They have our guy drive nearly 30 kilometers from the outskirts and have him take an exit ramp that has a street light at the end of the ramp. They have him get out and turn around, lifting his shirt and pulling up his pants legs to show he isn't armed, then instruct him to take the bags with the cash and walk down the arroyo to his left. Once he moves out of the light, we all hold our breath: this is the most dangerous part of the operation. After about 15 minutes he comes back to the truck and gets in. The lights come on and he turns back onto the toll road going back towards the city.

We are set up overlooking one of three roads that connect to the highway. We have been there for about two hours when the commander tells us to move up to where the cash had been dropped. We move up using the detector to zero in on the sight. We are getting real strong signatures on the device and finally locate the drop site. The two black bags that contained the cash are lying on the ground. We get out and I notice the discarded packaging of three or four space blankets. These are lightweight sheets with a metallic material that traps heat. They also make an excellent Faraday bag if you suspect someone has hidden a tracking device in your payday. The thieves would have watched the drop then moved forward and retrieved the cash, discarded the bags, then wrapped the loot in the blankets before putting it in their own.

The surveillance units that were dispersed far enough away, but close enough to monitor the area, had only detected two vehicles that had entered the area. One was dismissed as the vehicle contained a man and a woman who took a piss break then left. The other was an older model pickup with a shell on the back that hadn't come out on the road, but they found where it had gotten back on the highway from its tracks. We are scanning the drop site and the road to see if we can pick up their signal. This takes about an hour, but we are fairly sure that we can follow. The hope is that we can be on top of them when they release the victim.

Jeff, who came down that day, is back with our junior member, who has returned to the compound. The kid is upset and has every right to be, but he did, sort of, volunteer to do the drop. Apparently Reek and I have passed the scalawag mark and have been promoted to gold-plated assholes; and he wants more money next time. Next time? The gene pool once again confirms itself. He is a Type A personality, definitely.

We are crammed into the van. Reek, Mick, and I are taking turns operating the equipment. We are casting along the route until we pick up the signal again. It's weird because the signal gets real strong, then weak, then strong again and we are trying to interpret the results. Finally, I get a notion that the reason is because one of the gang is a smoker. Each time he opens the window our signal manages to escape. We keep following, trying to close in on them so that as soon as the victim is free and safe the state police can move in. They have a flying column with SWAT and support positioning themselves based on our location, and since they have an ID on the gang, positioning for raids on the houses where they are known to use. It's a long process and we are in the country trying not to be conspicuous yet stay on top of the signal.

Our driver is a state police officer, who is a perfect match for our band of rabid chimps. He listens to our bitching with interest, and occasionally makes a cryptic remark about having second thoughts about our having access to the extra guns he has brought along, since we don't have authority to be armed. Now anyone who knows us would hesitate to arm us and then lock us in close proximity to each other, but we had convinced him and his boss that if we ran into the banditos before the reaction force could get there, five guns in a firefight is a lot better than one.

He has brought four pistols and two submachine guns along, and of course there is the ongoing game of musical chairs every time we change operators, with the winner getting the seat the sub is parked under. Each time this looks more like a wrestling match accompanied by tripping, punching, and one time where the gun actually changes places.

The bloody machine takes all your concentration, so you are stuck in the most vulnerable position, since your eyes are glued to the screen with a hood over you and the laptop so it doesn't light up the interior. So if anything bad happens your first warning will be excited jabbering, followed by either rapid acceleration, or a 7G braking. We are switching over every 45 minutes, but it has been a long night. We are all dressed in comfortable clothing since we knew we were going to be sitting all night except for the few breaks to piss or change operators. So far, the machine works well, and we are getting better at deciphering what it is telling us. We on the other hand want to kill each other; the cop wants to kill us all.

In the middle of nowhere, where you shouldn't get a signal, my phone rings: it's the Beast casually inquiring about how things are going. I notice Mick toeing the sub out from under the front seat and moving it inch by inch back to his. Reek doesn't even turn around but tells Mick that if that gun isn't where he left it, that he does have a magnum handgun also, and that he will blow out his kneecap and leave him behind to be finished off, buying us time, so that he and I can escape. The officer has given up hope that he will be rescued in time.

Jeff and I chat for a few minutes and I get a stroke of genius: I make like it's the *Comandante* on the line now and start a one-sided conversation about how his boss is saying that they believe that there are nearly 50 *sicarios* in the gang and that they will probably be at the target area. The officer is talking on his commo link, obviously inquiring about validation of that news, and the two monkeys now are having a pulling match for the sub-gun and extra magazines while suggesting that we switch operators. Since I am in the bad seat, I will be part of the mauling when we switch. I tell the officer to pull over and we are going to switch. He turns and makes the universal screw loose motion, finger rotating to the head, but slams on the brakes and pulls to the side of the road. As soon as it nears stopping, I have taken the laptop, pushed it up on the stand, loosed my seat belt, have a hand on the door.

We roll to a stop and I am out like rocket and barely miss the behemoths' rush to get at the gun. I had to piss so bad I thought I could float. I finish and start back to the van. We are parked in the intersection of a T where an equally primitive road met the one we were on. It has been a bone-jarring last few hours and we don't know where the end point will be. We turn on the lights when we get to a highway, but we are following a trace signal and it keeps fading until we find the right direction. These guys obviously know every short cut back to their destination because we have been off the highway at least three times on country roads.

It's been another two hours and two shift changes. The boys have settled into a sullen standoff. The officer had hissed at me after the last piss stop to please not do anything more to rile up the simians, so I have been whiling away the time trying to remember the lyrics to "Pancho and Lefty" and making a crude attempt at translating them into Spanish.

It at least keeps me busy as we are jostled onto yet another dirt road going up into the foothills. It's pretty country, real old Mexico; we pass a few homesteads, but they are dark. The first tendrils of dawn are beginning to creep up from the East. We top a hill and the signal gets the strongest that we have seen all night. There is a small village, mostly dark but with a few lights of the early risers showing. There is but one street light and it is in front of a rambling block building. I have to pee again, so we stop, and Mick keeps calling out readings. I walk back and down the hill a ways and relieve myself. I want a cigarette, and my nose picks up the smell of one. I hope whoever it is has taken the precaution of cupping it and his face when he takes a drag.

I am hearing someone shouting up by the van. Whoever it is, he is shouting in Spanish and now the village dogs are beginning to take up a crescendo of barking. At least two roosters join in. As I get to the van the officer is standing next to it and makes a motion towards the front. I start in that direction. Mick intercepts me to tell me that the *Comandante* believes that the village is the target and is 20 to 30 minutes behind us. We are overlooking the village on the only road in or out. I move forward and there is Reek. The dawn is lighting up the scene. He is in his expensive running suit, with a magnum handgun tucked down the waist band of his pants, the sub by his side in his right hand.

He is shouting at the village in general, using very profane language, about how he has been cooped up for 36 hours with two *pendejos*, and that they have but one chance to surrender or he will gladly kill them all. Mick has gotten out of the van and is arming himself up, having shut the detector down to idling. He looks at me and rolls his eyes in the direction of the ogre, and hands me a handgun and two magazines with the comment that this is an interesting approach. The officer comes around the back and asks me pensively if Reek knows it will be at least 20 minutes until the cavalry gets here. I just shrug and start looking around for safer terrain if things get bad.

Reek finishes his rant. More and more lights are coming on as the village wakes up to see what the ruckus is. He moves past me, still swearing and getting into werewolf mode. He looks at the three of us and spits out that this is what we came for so we might as well get the dance going. That 20 minutes seems a long way off at this point. We leave the van, back away from the road and take up positions on both sides of the road on a cut. We have the high ground at least. The officer has added a shotgun to his arsenal, my two gorillas have taken the high ground on either side of the cut.

The village is awake enough that a few furtive figures dash from the houses headed for the block building. The officer is next to me and looks up and asks if Reek is always this way. I tell him mostly, that's why I wanted the guns. He looks at his watch hopefully. Just about the time the hombres in the village get enough *cajones* up and start to get a vehicle up our hill, a convoy of state police in vans with three military hummers comes roaring up. They rush the village and seal off every goat trail out. There are sporadic bursts of fire in and around the village. Two armored hummers pull up to the blockhouse and vomit two squads that rush the house and do a rapid door entry on both ends. There is shooting inside the house and in two other locations. We wind down and I tell the two commando heads that playtime is over, and they come scooting down from their perches in the rocks.

The *Comandante* comes up and we spend around 30 minutes waiting while they clean out the rats' nest and secure the area. More state and army units show up. We are bushed and we still have a ways to go before our job is done. We decide that the entertainment is over and tell the commander we are going to go get something to eat. He wants us to drive by a location near the city not far from where the victim was released. Not a simple stop. It takes three hours before we are done. We find signatures on a house that was near an industrial site with a power hammer. This was the sound that was in the background when they talked to the kidnappers. Additionally we find trash bags along the road next to a drainage ditch that they had used to haul the loot from the village back to the safe house. Inside are fast food containers, with Chinese food, which is what the victim was telling the police that the kidnappers had fed him; these later prove to have his and several others' fingerprints on them.

Good solid police work: these guys know their adversary and were quick to locate where they held the victim and—with our help—several houses that contained other members of the gang and portions of the ransom money.

By the time it is over we are drained and exhausted. We stop in the city, park in the Gigante parking lot and go to the restaurant and eat. We are all famished. We are joined by the commander and some of the team. The parking lot is full of holiday shoppers as it's close to Christmas. We are parked facing the next row of cars. As everyone is standing around I am shutting the machine down when the machine spikes through the top of the scale. I look around and notice a young man, with two older females, walk up and open the trunk on a late model Chevrolet Caprice classic, to place a number of packages inside. The commander comes over and I point it out and he talks into his radio. The outer units stop the car outside the lot and it turns out he has some of the ransom money on him. The car's plates were from Michigan.

It turns out that the gang had rushed out to spend the loot on Christmas. We leave one guy with the machine and go eat. When we spell him on the machine he has spotted three more cars that were hot and the police are rounding them up.

I am later able to link up with the Beast back at the hotel and he will take off tomorrow for LA. I will follow in a few days. It's good that he came down so he could babysit the family and leave Crap Game, our liaison, to do his magic.

All in all it was a successful contract, with the technology actually outperforming. The software needs to be tweaked because unless you have a hunter's instincts, it's hard to see what the data is telling you. I got a chance to ride in a stuffed can with two petulant simians while solving Schrödinger's equation in my head. Who could ask for a better life?

CHAPTER TWELVE

Nigeria with Snow

We have been back for a couple of months when we get a call from one of the most unusual and enigmatic people that we know. This is a man that we will form a lifetime friendship with. I had met him through his twin sister, who is a charming and wonderful addition to anyone's court. She and her brother share a history that is impressive by itself, and individually have polished their resumés closing commerce in a market that is fraught with danger.

Dr. B. and his sister had both been Freedom Riders in the sixties and have politics that are sometimes the direct opposite of mine, but I always treasure their input. Dr. B. speaks a dozen languages and has a depth of contacts that rival anyone's on the planet. He is a respected academic, writes papers for a plethora of professional groups and shares a sarcastic wit: one of his favorite sayings is that "Russia is just Nigeria with snow." He can be counted on for contacts and influence, and I hope in the end I didn't embarrass him.

It starts with him asking me if I could provide escorts for a group of Russians from ports all over Russia. This request seems to be tame and yet might be rewarding. I call the Beast and we work out a schedule so that we have someone to shepherd the flock. The group is in Los Angeles for two weeks and we are to link up with their senior guy.

Yuri, it turns out, is a jovial caricature of the kind of man one would imagine could run a port organization in the heart of Soviet Russia. He is gregarious and speaks excellent English. The rest of the crowd is almost cookie-cutter, with a few deviations for the somber and the truly demented. A good bunch, but they have two teams of handlers. The first one we meet is the State Department sheep dog, there to escort them through the itinerary. He is a droll, bland, usual-stamped copy of a junior foreign service officer, who I am sure would have rather been playing tennis on the mall, than be here babysitting the waterfront crowd.

The second is their political watchdog from whatever organ in the KGB makes sure their foreign delegations are minding the rules, and don't use the opportunity to skip and discover the joys of fast food and full store shelves. He is across the board dour, skeptical, argumentative, and a royal pain in the ass. Yuri must have a lot of juice back in the Motherland, because he treats him like the petulant little sneeze that he is, with no fear of reprisal.

We spend a pleasant couple of days riding in the bus with the group as they are shuttled to various photo ops with the politicians and glad handers. We are fortunate that we don't have to stay and listen to all the windbag presentations, and can slip out for the droning two-hour sessions then come back and pick them up again. The State Dweeb stays and takes copious notes, the KGB guy has gone through three notebooks and has carpal syndrome keeping up. Yuri and his merry band are bored senseless.

They have attended ladies' teas, toured port facilities, luncheons, dinners, and a host of other functions that would give me gas just at the mention of them. My instructions are to entertain them within reason, and I have been given a budget which so far is unused.

On Friday we stop and have an early dinner with them; the air has that electric twinge of energy that hints at something being afoot. Yuri pulls me and the Beast over to a table with most of the higher-ranking members of the theatre of the macabre. They are all jovial as the vodka chaser race has started. We are pulled into their happy group because we are quote "good guys" which makes the State guy go stiff. He starts to say something but Yuri waves him off imperially. He leans over to me and the entire table with the exception of the Beast, who is watching the tableau like a barn owl checking out a group of nice fat mice. He cocks an eyebrow at me. Yuri leans closer and starts, "Neek we have greatly appreciated your efforts for Dr. B. and you guys are real guys, so we can talk, *da*?"

I look up at Monsieur Owl and he chirps in, "They want hookers or strippers," then goes back into hunt mode.

Yuri now has my forearm locked in a huge ham and intensely continues. "*Da*! We have been to every boring event, have seen what we wanted to see, and put up with the game. Game is right?"

"Yes it is right," I reply.

"We want," he pauses and looks around over at the State functionary and the KGB guy, grips me a little tighter and says, "We want to see some large American breasts, dance, drink, and maybe meet some nice ladies." I look over at the twin towers of control and both of them look like someone goosed them with a Taser.

I tell Yuri, "So let me get this straight, you guys want to hang one on, go to a strip club, then go dancing, drinking, and hunting. In that order, right?"

He looks at me, his face effused with the glow of having found the Holy Grail, and almost shouts out, "Yes, yes my friend, yes this is exactly what we want." I ask him what about the two babysitters and he ignores it, says something in Russian, waves at the Beast and me and the entire mob breaks into applause and except the two fun spoilers.

He pulls me to my feet, and the entire mob sweep the Beast and me out of the restaurant and onto the waiting busses. The State guy is busy paying the bill and they have locked the KGB guy outside the door of the last bus.

The Beast and I know the perfect place in Anaheim. It's one of the classier gentlemen's clubs, but they are closed for some reason, so we opt for the comedy place in Costa Mesa.

The State guy comes running up and motions that he wants to talk to me outside. I get out and he pulls me away from the open door and goes into this long rant that this is wrong and approval would be needed, blah, blah, blah. I stop him and explain that is perfectly fine with me—the Russians seem to have made up their mind, but he would be welcome to tell them *nyet* and his rationale. Then I pat him on the shoulder and tell him that both of us would be happy to tell his relatives of his sacrifice. The KGB guy has finally threatened his way onto the second bus and we head for Costa Mesa. Luckily we know the owners so we have phoned ahead and they have set up a special section to keep us from the unwary and meet us with two of the biggest bouncers I have ever seen.

We get everybody situated and the Beast gets up with the owner and lays out the ground rules. This is not a bordello, hands off the ladies boys unless they tell you it is okay. They are ecstatic over the put-a-dollar-on-the-candy concept, but none have dollars. The budget takes a ding but we soon have enough one-dollar bills that the girls are going to have to turn it in after every dance so the bar can make change. At that point it becomes the rowdy, loud and boisterous jovial party that the Russians are famous for. The booze is flowing, dollars are slipping into garters and other coverings and everyone is having a good time. I can see the testosterone is starting to rise and Miller and I decide the next stop, which will need to happen before they turn mean as spit and start serious drinking.

The Beast and I have the perfect place to lead this band of vandals. There is a well-known hotel that overlooks the back bay and on top it has a night club that literally hops on a Friday night. The place is loaded with cougars, and widows of wealthy men that they married when they were in their twenties. The hubby has passed to his rewards and in the overall scheme of wealth distribution, they are loaded. Both groups are horny as alley cats. The male part of the crowd is mostly Persian gigolos, who landed here with the Pahlavis. With their polished repertoire and custom-tailored suits, they are the perfect victims of this crowd. The Persians will have a jolt of primal fear from their past, that of a Cossack raid fueled by lust and vodka.

We pull up out front and the busses disgorge the happy crew and we make our way to the lobby. The small crowd of normal people soon edge back away from the elevator where the Russians are shuttling up to heaven. We had gone up first, pulled the manager over and explained this was an official delegation from Russia and that they wanted to enjoy some of what Newport had to offer. We make arrangements for the bill, leaving the State guy to figure it out. I pass out a little pocket change to everyone and the party grows. It's pretty smooth at first, with the boys pairing up when and where they can with some cougar that's thanking her lucky stars for the break from being scammed by the Persians. Not all these guys are raving beauties, but they are definitely male. Even the State guy seems to have calmed down.

Trouble starts when some gigolo starts to bitch about the "rough crowd"; unfortunately the Russian he directed his remark to spoke fairly fluent Farsi and grabbed him by his junk and frog marched him over to the wall for a little face time. A couple other of the Persian crowd get roughed up and the bouncers are deciding if they like their job all that much, when I get Yuri and tell him to get everybody settled down and that we are leaving in 30 minutes for their hotel.

Some of the lucky ones have disappeared over the last few hours, and I suspect there is a run on rooms being paid for by their recently acquired amour. I mention this to Yuri, and he says that they all have his number if they get stranded. The State guy doesn't look all that well.

I notice that the KGB guy is really plastered and seems to have four new buddies who keep plying him with liquor. We are starting to drain off and down to the parking lot. The elevators are glass tubes that run down the outside of the building.

Yuri, the Beast and I are going down to get the busses organized when we notice the elevator next to us go by. There appears to be a beat down going on inside. It reaches the bottom floor before we do as we stop on the 12th floor to pick up a couple who evidently haven't heard that the Visigoths are in town.

When we get down we are herding the revelers onto the busses when we see a patrol car come up, then another. They disappear behind the hedges that border the parking lot. We wander over and there are six of the revelers holding up the KGB guy who looks like a lump factory. The cop asks who we are, and we shove the guy from State forward, who by now is ashen but surprisingly still has composure. The cop looks at the crowd and asks if they are all Russians; we tell him yes. He points at the victim and asks who he is and we tell him that he is the KGB watchdog and that his charges were administering a bit of social justice.

The officer looks at us and says, "Take it over behind the bushes next time, where the public doesn't feel inclined to light up the 911 switchboard." They watch and wait for us to get all aboard. They give us a jaunty wave as we pull out. The Russians are singing some sentimental song off key but with great gusto as we head back to Los Angeles. The Beast, the Russians, and I are pleased that their night of fun ended without fatalities or arrests. The Russians love us and tearfully explain what good fellows we are. We get them safely back to the hotel and tucked in. The State guy won't even look at us, doesn't even wave as we drive off.

The State Department send some sharp shooter to take over from the kid, who probably volunteered for the Congo rather than come back. This one acts as if the Beast and I are lepers and never tries to be alone with us. The next morning the

AWOL members arrive in a variety of Mercedes, Bentleys and other high-priced chariots; they have arranged for a barbecue on the beach for the lads, inviting some of their friends as added entertainment. We made contacts that would come in very handy later, got paid, had a good time, and Dr. B. was happy.

* * *

We are content with our lot as long as we are able to grab a project, right at the edge of financial disaster. It has nothing to do with our ability to market ourselves, which is shaky at best; but rather fate has some ulterior plan to keep us in the arena.

The USSR is going from glasnost to disintegration, and all the pundits are aglow with principles of governance and predictions of the golden age of free markets on one side and the sullen denial that communism was a failed ideology, on the left. We have been doing some work for the resource development crowd and are starting to get inquiries from folks that want to get in on the looting. Most of these we dismiss but there are a few that have at least a modicum of intent. I have been talking with Dr. B. since he is intimate with the situation. He knows most of the Moscow phone book, actually writes position papers for a number of respected journals. His information is as usual better than one would get from someone who collates data in the basement of the Post Office. I have a client that wants to explore the possibility of getting red oak out of Siberia in volume and others that are showing an interest in the area.

He is explaining the intricacies of that in terms of acquiring the concession, navigating the political structure, requirements for kilning to kill off the pests and prevent re-infestation, customs etc. He deftly slides in the carrot, with "As long as you may have to go there … " then launches into some tasks that me and my simian relatives might be able to do while we are there.

Russia? The land of the Rus and *Dr. Zhivago*? Long sleigh rides on a snow-covered plain, pulled by terrified horses, whilst being chased by wolves, next to a ravishing blonde?

This is the problem of being an adrenalin junkie. Your attachment to your mortal coil is fractured by the thought of the exotic and bizarre. We can't resist. We probably won't get rich but will most likely pick up other opportunities along the way. I'm hooked, which didn't take much effort on his part. I still have responsible paranoia so there is a thought seeded in the back of my mind that it is also entirely possible that he is hoping that I will end up in the gulag, and quit pestering him for assistance.

Of course, my first call is to Jeff. I explain the tasks, the money, and the timetable. The Penguini would be a nice addition, but he is on some health kick and most of his luggage is gym bags, besides which, he isn't answering his phone. We in turn discuss the trip with the government through an old contact and get advice on restrictions etc. There don't appear to be any but we are going to show up on someone's radar probably. At least they will know where we finally disappeared from.

The next week is all preparations: we will fly to London, spend a day or so then launch into Moscow on Aeroflot. We have been provided an interpreter and all-around contact who will meet us at the airport. The trip over is interesting. There is a smattering of dour government officials, a few business types, and a legion of lawyers, consultants, and carpetbaggers.

I get up to use the loo shortly after we reach altitude. I am almost to the rear before I start seeing people that look and dress like Russians, it is like a culture wall separates the two groups. I also discover that Russian commercial liners have a negative pressure plumbing system. After its third or fourth customer, the back of the plane reeks of its contents, and some chemical disinfectant that will remove paint with smell alone. The stewardesses are perky and efficient, supervised by an older, fiftyish woman with the mannerism of an East London char woman. She comes by a couple of times and gives us the once-over, usually to tell someone two rows back that "*Nyet*" means no in English.

We arrive in Moscow and our escort is waiting for us. He is a small moon-faced man in his fifties. He is wearing the ubiquitous blue serge suit and a London Fog rain coat. He ushers us through immigration and customs quickly and efficiently and we take the car he has provided, with a driver, to our hotel. The hotel is the Russian, at the foot of Red Square facing up to St. Basil's with its glittering gold domes. There is a legend that when all of Moscow's churches have gold domes, Russia will return to greatness. There seems to be a belief in it because new gilding is evident everywhere. We have a business suite for each of us. This hotel is where the illuminati from the provinces stay when they are in town and has a buffet that will give you gout just looking at it. There is also a nightclub, with floor shows run by the Armenian mob.

We are close to Arbat Street, so we take some free time to walk the square and ogle at the place we have only seen in news reels. Lenin's tomb is across the square and we manage to get inside after a wait in the line outside. Lenin looks better than he had in real life, so much so that you expect him to leap up and shout at the crowd about running-dog lackeys of the bourgeoisie, and run off howling to catch the Red Arrow to St. Petersburg.

Arbat is one big street market, where you can buy everything from militaria to paintings, heirlooms, or a MiG 27 with spare parts. It abuts the square. What is missing is restaurants: there seems to be only street vendors but the smell is mostly cabbage farts and moldy cardboard.

Sasha, our guide and majordomo, comes back with an itinerary of people that we have to see that will assist in accomplishing our objectives. I have to say we find the list impressive. Sasha had been the third secretary to the UN mission for a number of years, and spent his youth as a functionary at the Russian Embassy during the war. He is an unabashed Americophile; his memories of America buoy him in the new era of cooperation and openness. We keep a rein on him and explain that not all is rosy in paradise. But if you came from a culture where 11 people live in a two-room flat and you're saving your money to buy a hat, everything, even Harlem sounds bright and new.

The Beast is in his element. We have taken to going to the Penta, which is an upscale modern Western hotel, in the afternoon when we are free. The place is full of foreign businessmen; the rooms are all wired for video and audio, and the hookers get dropped off by their husbands when the shift starts. In those days an astrophysicist made the equivalent of 300 dollars a month so hooking on the side was survival. The women are stunning to say the least. They have a great bar and restaurant. Our days normally stretch well into the night depending on who we are meeting with.

Our primary contacts through Sasha are former members of the KGB and Politburo and the technocrats that run the various state organs and industrial blocs. One of our additional clients is a non-ferrous metal conglomerate who has interests in the commodities market. They have verticalized their operation and are starting to buy into mines, rail, shipping, and processing in order to get their metal out of the country. Our first chore is to find someone that can act as their legal and security arm in the countries they need to operate in. We have a contract, already drawn up with a cashier's check to start the relationship, once we get to the right people. It has a lot of zeros on it.

The client has bought entire blocks of buildings in Moscow and St. Petersburg and is in the process of completely gutting them and rebuilding them like the ones in Prague or Vienna or Berlin. These are huge rococo buildings, five stories high and stretching for a hundred meters on a side, encompassing a central courtyard. Usually there would be shops and restaurants on the ground floor, then offices on the next two with apartments on the top floors. Very efficient design. We visit a couple of sites in Moscow and a couple in St. Petersburg when we get there. The client wants to know why they are getting robbed blind and by whom because they are way over budget. We determine that their own partners are at fault.

The Russians are supposed to be paying the workers, half in rubles and half in hard currency. They aren't. They're giving the workers the rubles and buying themselves accounts in Switzerland and mistress-equipped chalets. We watch as one construction crew loads 40 Western toilets in the back of a dump truck, covers them with sand and drives off with them. Everything of value is being sold on the black market. We saw a few of the toilets and counters at an outdoor market two days later. We need someone with juice and a strong arm to counter this. Sasha finds us the perfect match.

We are on our way to see one of the senior figures in the KGB at his offices. We are shepherded to the top floor of the Lubyanka. Ah, Lubyanka, where Gary

Powers was interned, the drab edifice that bespoke of the dark side of the totalitarian regime. This is the gaping maw of the gulag, the doorway to the state-run prison system. It's all very, very intimidating. Yet here we are. The Beast is soaking up the ambience, I'm not sure that is healthy, he is already taking on a John le Carré look and demeanor.

We are ushered into the office of the director, which is a wood-paneled palatial office with a sitting room and padded red leather doors. Meeting us is the General and his aide. The General is a tall, urbane, courtly man with a beautiful baritone voice. His deputy fits all the descriptions of the soulless functionary, who would gleefully snatch off body parts as part of the interrogation. He is short and very muscular, with a round face and a florid complexion. We are offered seats and Sasha makes the introductions. I am watching the deputy and he is assessing our every word and motion as we begin.

The General speaks to me in German, saying that he understands that I speak the language and asks me if I would prefer that or English. Sasha had given him a long introduction in Russian telling him our purpose and directives. He switches to English, which he is also fluent in. By this time I am fascinated by the two of them.

We outline our plan and the intent of our clients and the fact that since the KGB is out of business until given a new charter, that there was an opportunity here to act as the problem-solving arm of a growing commercial enterprise. I also add that it might be possible to move some of their midlevel managers into a commercial company to take care of legal and security problems.

He tells us that they already have done such, in fact they are in a commercial company. They were already doing due diligence work for one of the largest law firms in the US, as well as a number of large multinationals. They have the structure and the right juice to iron out any wrinkles, stop the growing organized crime groups from interfering, and generally make the whole system run like a well-oiled machine.

We had done some of our homework with Sasha and with London. We knew about the law firm and what they were paying the Russians to do their due-dilly process. The Russians were charging them 30,000 US for each finished portfolio. The law firm was selling the same report to their customers for 10 to 20 times that amount. I bring this up towards the end of the meeting. They both get squinty eyed at this news and ask me if I am sure; we tell them they should use their resources and see if our information is correct.

By the time we are near the end, we have agreed that they will take on the commodities folks in the same type of agreement they have with the other large firms. The Beast and I are giving them suggestions on how to stick a skewer through the law firm and others that have been taking advantage of their low prices. To finish our meeting, I show them the bank draft that the firm is willing to give them as their retainer: it's large, very large. At that point we are in the fold, the deputy actually seems to mellow as we shake hands with the General.

He motions for us to sit down and says something into the intercom and a secretary comes in with another woman with trays of cucumbers, meats, olives, and cheeses, along with two bottles of vodka and six water glasses. The deputy pours a generous measure in each glass and we toast to our new relationship. We spend the next hour socializing. It's a pleasant afternoon.

As a kicker he pulls out a file, and jokingly says that he was at first hesitant to talk to me, a former member of Special Forces and of the Detachment in Berlin, which everyone knew were trained assassins. On the top is a picture of me and two others that had obviously been taken in Berlin several years ago. He does this all with a smile. So much for secrets. It could all have been a con, but the picture was real. I don't get to finger the file, but it is a nice prop. All in all we would find their approach and assessments on the mark and delivered with panache. Through this group we had access to the shifting and ever-changing structure of a society rewriting the rules as they went along. We have also set up meetings with the oil industry and with their tech cities. They have ventures that they want to pursue as well.

We attend a few social affairs with the KGB crowd. One in particular stands out. We had been invited to dinner at the Officers' Mess. The General and other directors are seated on the top tier of a fairly large hall, built on levels like a stair. Each section from the top down to the lowest is assigned to a specific organization in the state security. We as guests are sitting at the top level. The militia, which is the police force at that time, occupies the lowest tier.

Before we had come in and were seated, we had been in the foyer. I had been talking with the deputy in English, when I was approached by an American that was with the militia officers. He had walked up and in a break in our conversation asked me if I was with the embassy. I gave him my standard reply to ward off curiosity. I told him, "No, I'm Canadian eh." But this didn't suppress his curiosity. I recognized him as he was the star of a reality show featuring real cops in action. They were in town filming with the militia, cop action on the streets of Moscow. The guy is trying to pry information of who we are, and we politely tell him that we are here on business and that it is none of his business. The Beast tells him to look back at his escorts, who are standing there, aghast that their guest's manners might lead to repercussions from the star chamber. He says to Hollywood, "Do your friends look like they are happy you are here bothering us?" Before he can answer two burly militia officers sweep him up and back to their group, while a third apologizes profusely to the deputy. The deputy lets him prattle on for a moment with a stone face and dismisses him with a wave. The star of the show still doesn't understand that he almost stepped in the punji pit.

As we go inside and are sitting down Jeff and I are still chuckling at his faux pas. The Beast says, "I don't like that guy, did you see his hairline, it starts at his eyebrows, he looks like a bad werewolf movie." I have to agree with him, the boy has a definite mangy lupine look to him. Evgeni, the deputy, is explaining to the

General what had transpired. When we sit down he asks me about the guy. I tell him what I know, mostly that he is on the Hollywood train. He remarks that he remembers their getting permission to film and that the militia was in charge of him. He playfully mentions that since we don't like him he could arrange they spend a few hours in the basement of the building.

I am envisioning him and his entire crew sharing Gary Powers' old abode, and listening to the tapes of their whining, but the Beast sees what is swirling through my peanut, and looks over at me shaking his head. No. It was a pleasant thought. I remember years later seeing a rerun of that episode and how upset the star was that snitches got to keep whatever narcotics were seized. He probably didn't read the fine print. They got to keep it but had to use it before they left the building.

We also meet a rising star in the political arena: a smallish, well-muscled lieutenant colonel who is fluent in a couple of languages, and obviously well thought of, as most of the senior staff treat him with deference. His name was not important at the time. If we had known then where he was headed we probably would have spent a little more time getting to know him.

There are several personalities of this age group: personalities within the system, who are hungry for change and realize the vast resources that the security organs control. They have their own military, they control all the earth sciences, the manufacturing bases, their own research cities that aren't even on the map. It will later play out on the world stage that these are the new Russians. I remember we were having a discussion with the General and he told us that whatever came out of the chaos of Russia, it would not be democracy, nor would it be communism, but it would be distinctly Russian.

This is all too heady for either of us because we are still in the rapture of being in Russia and doing what we do. It is apparent we are being worked, but we are very careful to couch any comments that whatever we are about to regurgitate is our opinion not any adherence to policy. If they are preparing us for later blackmail, they are doing poor homework. We aren't important in any way. In fact, our government considers us in the same light as ruptured hemorrhoids. At this point we are just hoping that if they plan on using the honey trap on us, we get to pick.

We get to see some amazing things because of this connection with the KGB. One such trip is over to the state-owned petrol headquarters. We have a client that is interested in the oil and gas industry, so we attend a meeting with their upper echelon, followed by a tour of what they are doing. We enter a large room that has a glass wall on one side and a sealed airlock of glass for entry. Inside the room are five brand new Cray computers and a cube farm feeding them. They use an iris reader for entry with a palm print salinity reader and a key pad to get us through. I ask one of our escorts why he seemed so intent on the procedure. Was it because he keeps forgetting the codes, and the alarms go off and he gets shipped to a gulag?

I'm joking; he isn't, when he replies that it was the poisonous gas that fills our space if he screws up. Very Russian, *n'est-ce pas*?

They are digitalizing all their paper files and loading them on the computers. They have all the necessary equipment, most of which I am sure was on the denied list. I ask them about it and they say the oil companies provided them with the equipment under some special permit, something, something, dash bullshit. Glasnost at work, and the seven sisters are at the head of the pack.

The security organ has possession of all the geological data in the country, and in addition they send geologists everywhere. Russia has a mission. As a result they have information on every country they have been in since they got conned out of Alaska. It is a well-known fact that Russia has more information on the mineralization of Guinea than the Guineans have.

I am looking at some displays on a screen showing the east coast of Siberia, and our escort comes over and says something to a technician and a map of Alaska rolls up, showing ore concentrations. He points out one and asks me if I want the data on a gold property that assays at over an ounce. I tell him that it is probably already being worked. He snorts and says that it isn't, and that he can have satellite imagery from yesterday showing that it is still wilderness. I still remember the geo coordinates, and if I get another lifetime I might go visit.

Another little tidbit is that the Russian electronics industry disposes of all their scrap through one firm. Their electronics are still heavy in precious metals such as silver, platinum, and gold. The boards and components are shredded and containerized then sent to one firm in the Benelux for smelting. That company pays scrap prices to the Russians and damn little more. Most of the material comes out of Belorussia and the Ukraine, and the oligarchs of both have deep ties outside Mother Russia.

We need to go to St. Petersburg, to link up with some of the General's subordinates and review a couple of sites being used by one of our clients. It has been arranged that we take the Red Arrow express, the train to St. Petersburg. He tells us that his people will meet us at the train station. I ask him how we will recognize them, and he just smiles a wolf-like smile and assures me that we will immediately recognize his contact. We gather up Sasha that evening and go to the train station. We have taken to riding the Tube around Moscow, it's efficient and well run. They have the longest escalators in the world which seem to go deep underground to the various levels. These are sometimes a hundred meters long, packed with commuters. This had been one of Uncle Joe's postwar creations to celebrate the great Soviet. The platforms and passageways are beautiful with rococo reliefs and frescos, lots of gilt everywhere.

When we get to the train station that we will take to St. Petersburg, it looks like a scene from *Anna Karenina*. The train pulls up to the platform and the engine has bronze sculptures of the same flags you saw in the movie on the front, all in their vibrant colors. We have a cabin, so we find it and get settled in, with Sasha physically ejecting people that try to force their way in and take a seat. He explains that it is a common practice especially with pensioners, who feel that they sacrificed for the Motherland, ergo they should enjoy its benefits. It's a small wonder because I wander back to the general seating and it is a packed rendition of a boxcar, with people crammed in every space, chickens and at least one pig adding to the clamor.

It is here that we discover one of Sasha's habits that defies logic. I had heard of folks that drink their own urine as a detox procedure. He not only does it every evening and morning, but he keeps it in a kind of mason jar in his luggage. Other than this it's a magical trip in the moonlight and I fall asleep with the clacking of the rails. The Beast moves Sasha to the lower bunk so he can act as doorman, but I suspect that he didn't want him to have an accident with his jar in the middle of the night.

As dawn breaks, we are coming into St. Petersburg. We get ourselves freshened up and are standing at the window watching as the train pulls into the platform. We are discussing how and where we will meet our escorts with Sasha reassuring us that the General said they would meet us on the platform. As the steam clears, I am looking for someone with a sign that has our name, like the livery drivers at US airports. Jeff begins to laugh and pokes me in the ribs, and says, "I bet that guy in the leather jacket is our escort. You can't miss him." I look and almost shite myself. I am looking at my twin. He isn't vaguely like me. He is my exact doppelganger. We stare at each other through the glass. I can see he is as astounded as I am. His companion is laughing uproariously, so are the Beast and Sasha. We get off and he and I circle each other like two dogs checking each other out. Apparently, the General hadn't told him either.

We make all of our meetings promptly and efficiently, and in our off time they show us this magnificent city built by Peter the Great. It has a sad air of decay but is still magnificent in its façades and cobbled squares. It reminds me of an elegant lady whose beauty is starting to fade with time. We hit it off with both of our escorts and are finished in two days.

As a treat we decide to take a boat down the Volga on the weekend. Our ship is a hydrofoil with cabins. There is a proliferation of older gentlemen, with what appear to be their daughters, on board. After about 10 minutes you realize they aren't kin. Sasha tells us that these ships are called colloquially "floating bordellos," as the well-off take their mistresses on weekend jaunts.

On board there are a mass of Cossack officers in full regalia, heading for some parade or function. They are mostly on deck and are drunk as English lords. They are hindering our departure with their revelry. They are still partially on the dock and clogging up the gangway with their companions shouting insults and suggestions to the late comers.

A Russian policeman with the ubiquitous white baton starts shouting and cursing at them, telling them they are a disgrace, and that he will ship them all to Siberia and other empty threats. The Cossacks ignore him until he becomes a nuisance. He is a late middle-aged person obviously with a lot of dignity. They bodily pick him up and start passing him hand over hand up the gangway, all the while cheering him for his diligence and at the same time stripping him. First goes the baton, then his pants and tunic, until he has only his underwear as he passes up the line, across the deck and then overboard with a hearty *bon voyage* from the crowd. He surfaces and swims to shore and they toss his clothes back on the dock. We pull away as he stands there dripping and furious, shaking his fist at them and calling them hooligans.

This is our first contact with the Cossacks but not our last. Each regiment has a distinctive stripe down the pant leg denoting where they are from. I am convinced that they have a spare liver from their alcohol capacity. Once they discover we are American *Spetznaz* we are pulled into their embrace as comrades. My kidneys still hurt when I think of them.

CHAPTER THIRTEEN

Into the Hindu Kush with the Beast of the Baskervilles

We are back in Moscow and our next stop is Kazakhstan. We have a client that is interested in the oil and gas business that is burgeoning into smaller national units with the break-up of the Soviet Union. London has given us some interesting contacts as well as the lads from Lubyanka. We will see if we can get our client the relationships that he wants.

After our previous experience with the etiquette of Russians toasting success with copious amounts of vodka, I am hoping that the Kazakhs use kumis, fermented mare's milk: at least I will get some nutrition with the sclerosis. We have survived so far because we took an old-school suggestion and ate a spoonful of peanut butter before the drinking bouts, but I have been mildly buzzed for a week. We are taking Aeroflot to the capital city. Our airport is outside of Moscow and we leave early in the morning whilst it is still dark. The plane has some unique features such as you load your own luggage into the belly in bins with nets on the front, and one can access the hold by a spiral staircase inside. Once loaded we soon are airborne. It's cold, Russia cold; although it is the beginning of spring, winter still has its grip on the vastness. After a couple of hours we pass over the Urals. I am looking out the window at the vast countryside with no visible habitation. The forests stretch forever with the higher elevations still mantled in snow.

We are some five hours into the flight when there is a loud bang, then grinding, coming from one of the starboard engines. It's enough to wake the Beast up from where he has been floating on a sea of Ambien in dreamland, probably confessing his sins to the higher authority and blaming me for everything. I am staring out the window at what appear to be flames coming from the engine. There are more grinding noises and a flash of mist which was probably the fire control system dousing whatever is chewing itself to pieces.

"What was that?" He pops up erect and tries to stare out the window, but his seat belt is still snapped so he can't make it. I push him back and tell him, "It's nothing, I think we hit a pterodactyl, but the engine ate the remains." The cabin is abuzz with others that express the same concern as the Beast looks at me. "I'm not bullshitting here, what was that noise?"

"Do I look like I am the flight engineer? I'll give you a layman's explanation, big boom this side, smoke, fire, now big plane wounded. Anything else you want to know?" He looks at me sourly and I am getting ready for another zinger when there is a second boom slightly less threatening than the first which starts the hubbub anew. A stewardess comes down the aisle telling everyone to fasten their seat belt. Sasha hasn't even stirred, no doubt comatose in a sea of his own urine and brandy. We are descending in a spiral not rapidly enough to totally freak everyone out and enough that it is apparent the crew has some control over this steel tube full of borscht eaters and us. I look out the window. I don't see anything but snow-covered wilderness below. Not good. The Beast insists on switching seats. No problem. Let him narrate what he sees, it will keep him busy.

We are starting to hit turbulence as we descend. The Beast calls out that he sees lights below and it looks like a runway. There are no other lights. Sasha finally comes to and asks what is going on and the people across the aisle give him the situation in Russian. He translates that the engine blew a compressor, but we are making an emergency landing at a military airfield. I'm peering over the Beast and all I can see is the runway lights through the blowing snow. We are bouncing in the turbulence and are down to about six grand and I can make out hills and forested slopes and a flat spot with the runway. There is a huge Quonset hangar, open at both ends. We drop into the landing with the wheels screaming as they are lowered. We use up almost the entire runway since whatever it was reduces the use of the reverse thrusters. We come to a halt and taxi back to the open hangar which is too small to house the plane. There don't seem to be any other buildings.

As soon as we stop everyone is trying to get in the baggage compartment and soon emerge with heavy jackets, blankets, dried meats, loaves of bread, vodkas, kipper snacks etc. Evidently this exercise is normal for air travel in this region of the world. We go down and get our bags and take out heavy clothing, some booze, and our survival kits. When we come back the crews have unassed the bird and are getting in a tracked vehicle that appears out of nowhere. They drive off into the snow and disappear. I ask Sasha what's happening, and he explains that they will send someone to fix the bird and we will be on our way. We ask him how long and he says it might be a week. Where will we stay? We ask him; he says here on the plane. Nice.

He says there is a military post about 12 kilometers away, but they won't help other than medical emergencies. The Russians go about settling in for the experience, with the elan of people who take these incidents as a normal course of events.

There is one group that doesn't seem to be very happy about the turn of events. They are five, obvious Americans, who are towards the front. They are grousing about the cramped conditions, the coming cold, the abandonment by the crew, and a host of other bitches. The Beast and I avoid them like the plague. Before long the interior of the plane warmed by 130 bodies is actually pretty nice. Someone has broken out a balalaika, and the atmosphere is reminiscent of a barn party. The heavy reek of the Russian cigarettes sticks to everything. We can hear more bitching from the Americans in front. Sasha has gone forward to get something out of his bag. I am hoping it isn't his urine supply. He comes back to report that the Americans are a group of attorneys that were on the way to "assist" the new republic in forming a democratic government. The Beast hisses at him that if he told the group that we were American, we would leave him outside the plane overnight. Sasha looks stricken but vehemently denies that he has snitched us out.

The night passes and the next morning we make our way forward and down the stairs to the cargo hold and outside. Someone has managed to get the door to operate and people are going outside because the toilets are no longer functioning. No wonder with the cabbage and beet diet. There is yellow snow everywhere, but most have opted for an outhouse next to the hangar for the heavier loads. It is crisp and cold and the thought of using the facility brings back memories of the one we had on the farm when I was a kid. I still remember frost burns on my ass from that experience.

We look out at our surroundings and notice that there is a lake, frozen over, about a half a mile from the hangar. Around midmorning we see a horse-drawn troika with several people stop and start auguring holes in the ice. I know what this is. We grab Sasha and walk down to the lake and strike up a conversation. These are locals and they regularly fish the lake. While we are there, they pull three or four fish which resemble pike from the lake. We buy some black bread from them and return to the plane. There is a copse of woods at the edge of the lake, thick with pine and birch. There is plenty of dry wood and perfect for setting up camp. I'd rather be out here than in the plane. We dig out our survival kits and begin to build a camp with a lean-to, three sleeping platforms with pine boughs for bedding and a reflector wall for the fire pit. We are finished by midafternoon.

We have spoken to the locals who tell us that they have no objections to our using their holes for fishing and agree to return tomorrow with more supplies. Jeff and I break out our gear, pull out our fishing lines from the survival kits, and rig a couple in the holes. A couple of hours later we have four fat pike and have two of them cooking over the fire. Sasha is helping us by gathering wood, constantly praising our ingenuity. He takes the two remaining fish back to the plane and soon returns with some more bread and a tin of lard. We have a nice meal of pike and bread and settle in for the night.

It's warm and inviting in our makeshift camp and a couple of the Russians join us with a bottle of vodka. We drift off warm and comfortable in our little cocoon. It's better than a snow cave, and preferable to the rank confines of the plane with its cramped humanity and odors.

It's dawn when my bladder forces me awake, so I get up and move out to the latrine area that we have built 50 meters around the small hill that the trees are on. I am finishing and I hear faint noises from the lake, then closer in the forest that the copse is connected to. Now there is no mistaking the sounds, it is horses. I can hear their snorts and the jingle of harness. The Beast joins me and asks, "Is that the sled again?" It can't be because the sounds are in the forest. We stand there listening and Sasha comes up, his breath making fog as he walks up. It's lovely out and actually warm with a Chinook blowing gently. The sounds get closer and in the light we see a dozen horses with riders appear out of the shadows. They are military and Sasha grins like the cat in Alice in Wonderland. He gleefully exclaims, "Cossacks, my friends, Cossacks." He calls out and a voice answers, then we are surrounded by men on horseback.

They dismount and their leader, a lieutenant, comes forward and greets us. Sasha and he have a spirited conversation. They all seem to regard us with wonder. I am fascinated by their horses and uniforms. It's a scene right out of time. They are resplendent in their kit. The lieutenant speaks rudimentary English. Our area has happy smells of horses, which are stamping their feet. The whole scene is magical. We soon realize that the bush telegraph has been working overtime.

They are a patrol from the nearby military post who come regularly to the lake to trade with the fishermen and patrol the area. They had been sent to check on the plane and its inhabitants and they tell Sasha that a tracked vehicle will be coming with an APU unit to fire up the plane's electrical system and a sump truck to drain the toilets. The entire troop is looking over our camp and beaming in approval. Sasha tells them we are former *Spetznaz*, which sets off a round of back slapping and comradely hugs. After an hour they move off towards the plane.

Soon after they ride off through the forest a track vehicle pulling a generator unit and sump tank arrives, with two more soldiers who soon have the plane's systems working. Sasha tells us that the military will have a repair crew there tomorrow.

Late in the afternoon we see five figures making their way towards our camp from the plane. It's the lawyers. They have wrapped scarves and articles of clothing around their Italian loafers and look properly miserable. They are bundled up in expensive overcoats and most of what was in their luggage.

They hail us and move forward. We need this lot like we need frostbite. They have a tale of woe about not having much to eat and the conditions, and a host of other bitches about being robbed for their valuables by their fellow travelers for a few tins of food. The group are eyeing the fish strips and our hoard of food. One of the Cossack left us sugar and black tea and we are enjoying some out of cast-off tins that we boiled water in. It tastes slightly of the former contents but is steaming hot and sweet. The shortest one starts this speech about how we are all Americans and in the spirit of such that we should share our bounty.

They look wretched, but it's all blarney, they have that dumpy look of people that could afford to miss a few meals for the good of their waistline and humanity. The Beast looks at them and starts in, "You got any money?" They look stricken and shuffle from foot to foot. They are incredulous that we would demand money and say so in indignation, but neither of us has pity for these idiots. One of them pulls out his wallet and fishes a credit card out and waves it at us and starts to give a speech about what brigands we are to take money from them when they need charity. The Beast waves a hand to shut him up and tells them, "Do we look like we have a machine here to process your card? The only thing that is good for is to scrape your ass and maybe start a fire. You have any cash?" Apparently, they had used all the cash they had on them on the proletariat fleecing inside the plane. He asks them if they have any watches or jewelry which starts off another round of whining. One of them yanks off a wristwatch and we give them some fish. I add to the conversation by asking them rhetorically that if I walked into their offices needing legal help, would they give it for free? This is the heart of the free market system, supply and demand. They leave with comments about telling the authorities about us when we get to Alma Ata. We give the watch to Sasha who accepts it nervously. Screw them, they are here to feed off the carcass of the disintegrating government. They deserve this and more.

Sure as sunrise, the military shows up again with more food and some vodka, and while we are sitting around our fire with the horses picketed nearby, another group arrives with mechanics, and starts working on the plane. They have also brought a field mess with boiling hot soup, which they ladle out to the passengers. We see the legal team at the back of the line giving us dark looks as they shuffle forward. We are trading some trinkets like lapel pins and articles of clothing. They mount up and move off after telling Sasha that the plane would probably be ready by tomorrow afternoon. We spend a comfortable night in our forest chalet and get ready to leave when the plane is ready. By noon the following day we have moved back to the plane.

The Beast and I have left a liquor flask at the camp site with the Special Forces crest etched on it for our benefactors. The crew arrives and a short time later they fire up the bird. The Cossacks have returned and are lined up next to the hangar and give us a wave, then chase the plane alongside at a gallop as it taxis out to the runway. If you replace the Kalashnikovs with 1901 Mosin Nagants, you would have an idea of what the Bolsheviks would have seen on Red Square at the start of the Revolution. They are magnificent. We lift off and are soon airborne.

We arrive in Alma Ata, currently in the process of changing its name to Almaty, and process through immigration with ease. We go through the Deputy line, reserved for officials and VIPs, and are met by an imposing figure, who though in civilian clothes commands attention from everyone in the building. The guy has all the earmarks of military and is in his mid-forties. Although he is obviously a native Kazakh he is about twice the average size.

We pass the lawyer twinks who are stuck back in the mob waiting for the two bored immigration officers who are processing the rest of the passengers and what's left of two other flights. They glare at us with malevolence. I grab Sasha's arm and hold it up displaying his new watch for them. One of them at least has the balls to flip us the finger. As we go out the exit, I spy their escort. He is a typical low-level embassy employee, all starched and looking like he just finished grad school. He is trying to dress like a Russian, but Americans just stand out. We have a rube look too I guess. We are soon out the door and ushered into a waiting car which takes us to our hotel.

I am ready for a hot shower and a good meal; so is the Beast. I can hear his stomach next to me, growling and bubbling. We arrive after a spin through the city. The hotel is right downtown, one of the new buildings, with a huge annex attached to it. It is a casino, which is shaped like a yurt and gleaming white. The sign over the casino is the Sahara same as the Vegas casino where the Soldier of Fortune conventions are held. How freaky is that?

We get settled in and go up to our rooms, agreeing to meet in an hour downstairs in the restaurant. It's a new hotel and done in western décor and plumbing. I take a long hot shower, shave, and change clothes. I feel like a new man. As I am leaving my room Jeff comes out of his and we take the elevator to the lobby. The hotel is posh yet comfortable. The restaurant is circular with murals around the crown of its domed ceiling. The food is a mixture of Western dishes and traditional. It is delicious, we devour it like wolves and are soon finished. The three of us are having tea, and Sasha tells us that we have a car and driver at our disposal, courtesy of our hosts. We decide to take a tour of the city since our first meetings will be in the morning.

There is new construction going on everywhere, mostly Turkish construction companies. There are high rises and cranes spread out showing a building boom that is just starting. We stop and go into one of the newest, which is a gleaming edifice rising up to the sky. It's a five-star, swank place, the lobby and grounds are buzzing with people of a dozen different nationalities. The city sits in sight of the snow-capped Hindu Kush off in the distance and has the climate of Denver. There are a number of parks and remnants of both the czars and Genghis Khan in the architecture, and many churches, mostly boarded up but a few are being refurbished.

All along the road there are gas lines on top of the ground sitting on cement stanchions. When they come to a driveway or a street crossing the pipes go up, then over, then back down and continue on their way. The guy who is driving us is a maniac, he is swerving through the traffic talking to Sasha occasionally and peeking at us in the mirror, then smiling with a 24-carat gold smile. The Beast has noticed the pipeline "tragedy waiting to happen" as well: we keep unconsciously leaning away as we careen around corners with the rest of the traffic narrowly avoiding rupturing the pipes. There are dents and dings on some of the pipes and some sections that look newer, so apparently our fears are not groundless.

We visit an outdoor market at Sasha's insistence, buy some brandy that the area is famous for, fruits, cheese, and I snag some chocolate. It turned out to have been packaged when Uncle Joe ran the country. Pretty dry. Sasha wants us to try mare's milk when we are in the meat and dairy end, which is booth upon booth of meats etc. laid out with big fly strips waving from the horizontal poles like Mongol banners. The Beast is stopped and staring at the centerpiece, which is a skinned horse's head that appears to be half smoked. The vendor occasionally reaches up and flenses a strip for a customer. He looks at me and queries, "What are the chances that's Trigger?" Sasha looks at the two of us perplexed until we tell him who Trigger was. The mare's milk is delicious, with a kind of salty taste. It is a beautiful spring day, so we walk through a couple of parks, viewing the Stalinist sculptures, glorifying the defeat of the Germans. They are cast from the remnants of German guns, vehicles and helmets. We even do a few tourist shots with a group of young soldiers. We finish up around four in the afternoon. Our appointments start tomorrow morning.

When we return to the hotel, the manager pulls Sasha aside and informs him that someone from the embassy had been inquiring about any new Americans checking in. He had told them we were only oil workers from America and that we had been here for over a month. It's probably the indignant and vindictive crowd from the legal profession stirring the waters.

Our contacts are first rate. We go to the offices of one of the most powerful men in the country. He is a parliamentarian whose official title is the Chairman for Internal Defense and Security. Basically, everyone that carries a gun in the country answers to him. His name is Bulat. When we get to his offices, the man that met us at the airport is sitting in the office with him. We are introduced to him as well, his name is Rustum and he shakes our hand as we all sit down for the first of what will be a series of meetings. I was right about Rustum's background. They both were KGB general officers. In fact, Bulat had been Rustum's commanding officer in Afghanistan and in Berlin. Rustum is *Spetznaz*, in fact he had been in command of all *Spetznaz* in Afghanistan during the war. Now, one is a major player in the Nazarbayev government and the other is retired and developing a number of concerns in the private sector. Between the two we have the perfect partners.

Our initial meeting lays out what we need to accomplish; with their assent, we map out a plan of who we need to see and when. Both men are impressive both physically and in their resumes. They also come from the same clan. The Kazakhs look nearly identical to our Native Americans, kind of like the Northern Cheyenne or the Utes. One can see the land bridge in their faces. They have the same hair, skin complexion, and fierceness. Over the weeks and months, we become good friends and comrades. The General had evidently told Bulat about my being in Berlin with the Detachment because he kids me about it occasionally as the time passes; that had we met then, our relationship would have been quite different. One of his jobs in Berlin had been keeping track of our small unit. They knew a lot more about us than those weenies at the Pentagon suspected. The welcome mat is out for the Beast and me, and we are given access to a number of opportunities, in return for our help on some of the projects that they have on the books.

The majority of foreigners in the city are connected to the oil industry. If you go to the glitzy hotels to the north of us, you can't throw a rock without hitting someone from the oil patch. The casinos there do a lively business relieving customers of their new-found wealth. The one with the most Texans has a restaurant that serves steak in every manner found on the planet. We are mostly insulated from that crowd and are the only two Americans in our hotel.

I am in my room on the third day, right after dinner, when there is a knock on my door. I go to the door and open it. There is a beautiful woman standing there with what appears to be a bottle of Cognac. I am thinking that our hosts have arranged for entertainment, and politely tell her in stilted English that I am sorry, but she has the wrong room.

She cocks an eyebrow at me and replies in prefect American English that she isn't a gift, but rather the owner of the hotel, and asks me to call the Beast's room and have him join us in my sitting room. I call the Beast, tell him that there is a hot redhead in my room with a bottle of cognac, and ask him if he wants to join us. His reply is that he doesn't do threesomes, he has enough problems with twosomes. I explain that it is the owner of the hotel and her rendition of the welcome wagon.

The woman is in her mid-forties and is a real looker. She is smart, well educated, and entrenched in the new oil-patch economy. Chevron had signed a concession for a field in the west of the country called the Tengiz and in the stroke of a pen had doubled their worldwide reserves. They were going full bore with the national oil company and a host of the other sisters, to start pumping the black stuff to market. That's where all the gas lines that fed the city came from.

She had been the mistress of the guy from Chevron who was the progenitor of the acquisition. He had developed terminal cancer and eventually died. She had been at his side and thus knew all the players. She had borrowed money from a Turkish bank, built the hotel and casino, and was one of the wealthiest people in the new country. She had paid off the loans in a year. The casino was a cash cow. She has come by because other than the embassy crowd, we are the only Americans in the capital. Within a year, however, you won't be able to throw a cow patty without hitting a Texan from the oil patch. She is witty and knows everyone and we become friends. We also find out that she had run a very expensive escort service in Denver when she met the love of her life. It is a long, enjoyable night with a woman who has the intellect of Plato and the savvy of a street fighter. A truly remarkable person.

One of the projects that we become involved with, in conjunction with our client, involves the two players we were working with. There is a backlog and bottleneck in transport between China and Kazakhstan. This is because the Russian rails are of a different gauge than the Chinese. This was deliberate, so neither nation could load several divisions on trains and cross into the other in the event of war, but now it is an impediment to commerce. Thousands of tons of material and equipment are getting stuck until they lift the cars, slide the requisite undercarriage under and send it on. The obvious answer is that they need a trucking company. There are a few Freightliners and Magirus Deutz and Volvo semis and they are the pride of their owners, all decked out in gleaming paint, with garish light displays. But you couldn't buy enough and get them in service to solve the problem.

One day, Rustum comes over and wants to know if we would like to go on a helicopter tour with him. We go out to the military base, load on a Mil 8, and fly over the countryside to look at the border and the logjam at the ports. While we are flying around the Beast looks down and asks Rustum about a large truck park of vehicles that are Kamaz tractors with stake-bed trailers. They are all brand new, just sitting there. Rustum explains that they had been ordered by the Iranians who had paid a substantial down on them for export to Iran. Relations had cooled between the Kazakhs and the Assatollahs, and the Iranians had defaulted on the deal. We look at each other: the answer is right there. Free trucks. Form a transport company and haul the goods between the two countries and charge by the kilometer ton. That's the beauty of being on the edge of new nation; solutions are greased by expediency. They have a lively business going within a month with goods getting moved both ways. One of our clients is in the cold mix asphalt business and we were trying to get them a contract to repair and build highways to expand into the oil patch. The increase in road projects would more than double the prospects available to the client.

Bulat takes us one day to the border and says we are going to have lunch. He and Rustum and Bulat's escorts go with us. We haven't time to even mutter an objection before we are on the Chinese side of the border. We go to this hotel complex perched above the rail port in the hills and meet with Bulat's counterpart from the Chinese side who also shows up with his escorts. They toast each other's success to the point that I strongly suspect that the Chinese have partial interest in the trucking business. They all treat the Beast and me with respect, but we are certain that if the embassy hears about us we will have problems from the stiff-necked crowd, or the cockroach crowd will want us to wear a wire. Neither possibility appeals to me.

* * *

The capital is a bustling vibrant place with the old bureaucrats becoming the new rulers. Under the Soviets, every important position had either an ethnic Russian as the minister with a Kazakh deputy, or vice versa. The collapse of the power strings from Moscow didn't cause the chaos that many of the smaller former Soviet nations were experiencing, because of this relationship. The move into a free-market economy is relatively seamless and we are there at the very beginnings. The country is a cross between measured development and scenes reminiscent of opening day during the Oklahoma land rush. The only thing missing is the Irish. That doesn't last: long soon there is a decent Irish establishment run by the appropriate hard-luck story from the great diaspora.

There is also a first-rate German *Gasthaus* which is a favorite place for lunch for the business crowd and parliamentarians. We go there with Bulat and Rustum one day and they are giving us the rundown on the place. Great food, good service, and the place is wired for sound. I ask Rustum about it, and he says that some of the bugs are theirs and some are the owner's.

We are introduced to the owner at some point. He is a genial man in his mid-forties. I keep getting the niggling that I know him from somewhere. It isn't until later that I put it together. The last time I had seen him was at some function at NATO headquarters. At that time, he had been wearing the uniform of a commander in the German navy. This place reminds me of Berlin but with a lot more activity. Everyone is here trying to make their fortune. The place hasn't yet been cluttered up with attorneys and accountants, and the sleek reptilians from the consulting houses. Business is being done at breakneck pace. We are offered a gas field next to the Chinese border and turn it down. I know squat about gas and oil exploration, except those ads in the eighties trying to pry the elderly out of their savings to invest in gas leases, being hyped by boiler rooms in Newport Beach. We could have flipped it I guess since we had the patronage of the most powerful clans, but hanging out with the buyers of such a deal would probably leave you mentally and morally syphilitic.

The weeks are full of meetings and short travels; we get to see a lot of the country. We fly with Rustum over an entire valley that is packed bumper to bumper with armored vehicles and personnel carriers, artillery and trucks, all pickled and sheathed in storage. This is the Soviet version of Reforger, in the West. In the event that the Motherland was attacked from its soft underbelly, they would load the First, Fifth, and Tenth Guards Armies on trains, and ship them south. When they arrive, they would unpack and fuel up, load the ammo from the vast dumps nearby, and do a left turn into the invaders. Later the Fedayeen from the 'stans would come across and raid these dumps for ammo and weapons.

We are invited to go with Rustum and Bulat to a traditional gathering of their clan. Without the cars and ubiquitous Soviet version of an SUV, it could have been a snapshot of when the Great Khan and his Horde were wandering the steppes. We drive for three hours out into the emptiness, to a cluster of yurts and larger tents with the Mongol-looking poles with banners. Some of the people are in traditional costumes, and some in civilian togs. There is a gaiety about the gathering and there are horses everywhere as well as equestrian events such as races, snatching chickens buried neck deep from horseback, and athletics that defy the imagination, all fueled by alcohol.

We are sitting around inside a tent with some of the luminaries and they are serving traditional dishes. I notice that the Beast has cornered a salver of what looks like large purple grapes, which he is popping in his mouth and gulping down. I ask Bulat where the grapes come from, as the Beast looks up from his gluttony. Bulat just snorts and congratulates the Beast about not being a squeamish Westerner, then adds that the little party favors are actually sheep's testicles, considered a delicacy. The resulting projectile vomiting from the Beast over his shoulder and out a slit in the tent, made him a legend. Even I was impressed.

One last weekend before we leave, and we go up into the hills to a hunting lodge with the lads and our driver. Its full spring now and the hills are alive with greenery, the streams gushing with snow melt from the Kush. It's a magical time. We stop along the way to wait for the rest to catch up before we turn off the twisting road to the top where the lodge is. I get out to stretch my legs, and the Beast goes sniffing around to his own pursuits. We will have about an hour wait so it's a lazy respite.

Our driver goes to the boot and pulls out a small spade and a pan. I know what it is for. I've used the same before: he is going to pan the stream for gold. Everyone here has gold teeth. During the Soviet times people would pan for gold and give it to the dentists who would melt it down and make you a crown. The normal material was stainless steel, like the villain in one of the Bond movies, but here you got a golden display. The more gold, the more important you were.

We meet one Canadian group that has bought a tailings field to an old lead mine. It is probably one square mile of 50-foot high and 100-foot wide rows of windrowed tailings. During the Soviet era, the miners had stumbled onto a gold seam and rather than report it to Moscow, had taken what they could reasonably hide, and dumped the rest with the lead tailings. The Canadians are now processing it and will pay off their investment in the first 12 months. Apparently, one of the former miners had emigrated to Canada in the heady fall of the Union, and had fallen into the hands of some astute fellows from Edmonton with his tale of gold in the tailings. There hasn't been a more opportunistic group of Anglo-Saxon speakers, since Harold the Bluetooth was making house calls in North Umbria. The Canadians are making money like bandits while our state department is publishing position papers to deter Americans getting in on the feed, so they won't have to actually do their job.

Our pleasant trip ends with the task of visiting Siberia to look at a timber project. We leave our guests with regret since they have been good comrades and the best of contacts. Siberia is a drab and muddy place when we get there. They have swarms of mosquitoes that can carry off a car. Everyone either smells of some noxious native prophylactic, made of tree bark and rotting carcasses, or wears heavy netting. We spend three days and gratefully embark for Moscow. Once you get away from the ports, Siberia is much like where I grew up in Minnesota, with the same mosquito population.

We spend a few days wrapping up affairs and sending a final report back to London. Our flight will leave from Moscow and travel over the pole to land in San Francisco. When we board we find a deputation of cops from the US that had been on an exchange with the Russians, a dance troupe of gypsies going to San Francisco on a cultural exchange, and their KGB escort trying to be inconspicuous. We aren't in the air 30 minutes before the gypsies break out their musical instruments and the booze, and the party is on. I fall asleep around hour four and remember being jostled once by the seats in front of us, then rudely awakened a couple of hours later by a fist fight that falls into our row. The party never stops. We land in San Francisco 19 hours later. As we are exiting the plane, I see the KGB guy covered up with his own jacket and apparently asleep. I look closer and realize that he is beaten to a pulp and unconscious. Some of the gypsies had recognized him from an incident before where he had beaten a couple of the lads while they were wearing handcuffs. They had taken him down in the cargo hold and tuned him up over several hours. We don't want to get involved so we leave him to his fate.

A few months later Bulat visits the US and the Beast throws a big banquet for him at the 1789 restaurant in Georgetown. He has oil executives and high-level military at the soirée and introduces Bulat to quite a few people who later play key roles in the growth of Kazakhstan as an independent state. Happy to help.

CHAPTER FOURTEEN

Bosnia by Bus

In 1995 the Balkans are in chaos, and as everyone knows, in chaos there is opportunity. Watching Bosnian women and children die every evening on CNN, in a hail of mortar rounds falling on the central market from the surrounding hills, is becoming a national pastime, but unfortunately it is also giving Miller ideas; which is never a safe thing. He calls and we agree to meet at Silky Sullivan's, I should have known better but I am a sucker for the outlandish. We order a couple of drinks, ouzo for me, scotch for Miller, and he outlines his plan. "What the Bosnians need is an elite commando unit to go up into those hills and dispatch a few mortar crews: throats slit, severed heads left on sticks … you know, deterrence."

"Are you out of your fucking mind?" My initial reaction is less than favorable. "First, where are you going to get the money to go to Bosnia and start a commando school, it will cost a couple of million bucks at least."

"No problem," he replies, "the Bosnians are Muslim, we can get the money in Saudi Arabia, I have already arranged some meetings with people I met during Desert Storm who have all the connections. And," he continues smugly, "I have the money for the trip to Saudi Arabia."

At this point he babbles some semi-intelligible story involving the Korean American Navy Cooperation League or some such and a retired general and Hyundai, yada, yada, yada. I quickly lose interest. The bottom line was that he is carrying 25 grand in cash. Ouzo is hampering my ability to follow his narrative, but it is also making me much more amenable. "Sure," I find myself saying. "What the hell, what could possibly go wrong?"

First stop London, aahhh London, always a treat to venture into the decaying heart of the Empire. Miller's connections prove as good as he said they were, and we have visas for Saudi Arabia (not always an easy procurement) on the second afternoon in town. Next stop Riyadh.

We fly into Riyadh and into heat, stifling heat, heat so hot it crushes your chest and makes your shoes melt into the pavement. It is fucking hot! We check into a very nice hotel though, with blessedly efficient air conditioning, and that night we go to dinner at the home of Khalid, one of Miller's Gulf War buddies. He reminds me of Anthony Quinn in *Lawrence of Arabia*; the Bani-Sadar-are-dogs-and-should-remember-it kind of attitude. The dinner is served *outside* on the veranda. Of course, since it is after dark it is only a balmy 98 degrees.

We explain our idea to the assembled members of the Saudi family, and they *love* it. Now Khalid is the guy the Beast helped bring a quad 12.7mm Soviet anti-aircraft gun back from Kuwait after the war to mount on the roof of this very house, so I get the feeling this crew aren't all playing with a full deck. But they are rich!

As it happens Khalid's father is somewhat of an Islamic scholar and is friends with a young scholar from Bosnia who is studying in Saudi Arabia. Ismail is currently studying in Jeddah, so the next morning we grab a puddle jumper over to Jeddah for a sit-down.

The kid is even more excited about the project than Khalid's family, and he knows just who to talk to. At this point in time, most assistance to the Bosnians is coming from Iran, and all of the coordination is handled in Istanbul. The head of this effort, a close personal friend of President Alija Izetbegovic, is a medical doctor of Sudanese origin.

Well break my heart. Moonlight on the Bosporus, drinking raki in small cafes under the shadow of Hagia Sophia, all that James Bond shit. And an average September temperature of 74.5 degrees. Let's go!

* * *

The three of us fly into Istanbul and Ismail leads us out into a fairly disreputable-looking suburb. This is not tourist Istanbul; this is more like the Kasbah. Distrustful faces with postures indicating hidden weaponry, shuttered doors and windows, packs of semi-feral dogs. We meet the Doctor on a second-floor balcony, and it is obvious from the start that he thinks Ismail is a dupe who has stupidly dragged two CIA agents into his little world. We explain Miller's plan and are met with studied indifference. Despite the enthusiasm of Ismail, and the promise of two million dollars from Khalid's group for the project, the Doctor is skeptical to say the least.

He doesn't kill us, which is at least a relief, but he doesn't exactly welcome us either. We are told that in order to go to Sarajevo and pitch our project we will need clearance from the Minister of Defense, so just go back to our hotel and he'll be in touch. Next morning, we meet with one of the Doctor's entourage in the hotel lobby with another gentleman, quite obviously some kind of Turkish intelligence officer, and get the third degree. He asks about who we work for, "Nobody," what our religion is, "Nothing much," and insists over and over again that we must be CIA: "No we're not," "Yes you are," "No we're not," "Yes you are," ad nauseam. The Beast finally asks "If I can get a letter from the CIA saying we don't work for them would that do?" He tells us we will just have to wait.

If you have to wait, Istanbul is definitely the place to do it. We spend eight days visiting the Topkapi Museum, touring the Sultan's old palace, the huge indoor-outdoor market where you can find anything from rare spices, to gray market Levi's, to hashish (at least that's what we heard). Evenings at the Zuni Bar at the base of the intercontinental bridge where the women are so beautiful it's like every night is a Victoria's Secret fashion show afterparty.

Finally, somebody must have decided what the hell, let these two idiots come to Sarajevo and we'll see what they've got, because we get the green light. Next morning we fly to Marseilles and our old stomping ground at the Hotel Le Rhul. Sadly, we can't link up with Carlos because he is up in Paris running his new company: a spy gadget store called CIA-KGB, SA. Carlos was never much of a one for subtlety.

We board a train in Marseilles for Zagreb by way of Venice and Trieste, which immediately breaks down and costs us a full day. We get to Venice and decide to spend the night and get a good meal: lobster, mussels, all the fixings, fine wine, sort of like condemned men having their last meal. How appropriate.

The train to Zagreb is a night train, so we leave Venice about 5p.m. and stop briefly in Trieste. By the time we get to the Italian-Slovenian border we are sound asleep on our benches. Suddenly very large women dressed in old Soviet-style greatcoats and wielding powerful halogen flashlights are blinding us and screaming at us, "PASSPORTS, PASSPORTS," while our sleep-befuddled minds grope to comprehend what is happening. We manage to find the correct paperwork and they stamp it and go away.

We arrive in Zagreb with the dawn. Lovely city, reminds me a lot of Munich; there is something very Germanic about the neatness and order on display. Well the Croatian regime were notorious Nazi sympathizers after all, something we soon learned the other Balkan people have still not forgiven them for.

Next stop on our little odyssey is Split, a seaside town in southern Croatia and the current headquarters of UNPROFOR, the completely useless United Nations "peace keeping force." Of course, there is no actual peace keeping going on, but they have to find some way to waste all the money they get from the foolish US taxpayer. The only way to Split is a bus that night, so we spend another night on questionable Euro-transport, on roads barely wide enough for the bus that winds along cliffs so you can look out the window straight down several hundred feet to the azure Adriatic frothing below.

We arrive in Split in the morning, it has now been three nights of very little sleep and Miller is getting noticeably surly. Cursing and mumbling to himself, pushing people off the sidewalk, this boy has to get some sleep and so do I, but we are on a mission.

Our first thought is to rent an armored car and drive to Sarajevo, but of course the world's press has driven the rental cost of the limited number of vehicles up to the ridiculous. "2,500 US dollars a day for an armored Land Rover, are you frikkin' kidding me, we are not CNN."

Left to find a cheaper alternative we discover gypsy buses depart the local depot at about 5a.m. and go somewhere vaguely near Sarajevo. They take cash and they are cheap! So, not really knowing our final destination, we climb on a bus after a few hours (a very few hours) of sleep and head off to points unknown in the Balkan war zone. Once again, I am asking myself, "Why did I listen to Miller, shouldn't I know better by now?"

Of course after about an hour the bus breaks down. Actually, it gets a flat tire and the process of changing it takes about an hour. Locals come out of their houses to watch and we meet a couple from Australia. Apparently, there is a fair size Croatian community in Australia and a lot of them came home to help their families through the "troubles."

Back on the bus, winding through the hills and valleys of southwest Croatia and eventually crossing the invisible border into Bosnia-Herzegovina, we are completely lost. We can only tell we are getting closer to the war zone because the damage to the towns and villages we are passing through keeps getting worse. Eventually they are just mounds of gray stone with nothing that even resembles a building left. Worse, the bridges have all been blown so the bus has to slip and slide its way down recently excavated dirt roads to river level and cross on very rickety looking Baily bridges that have been thrown together.

We stop for lunch somewhere. May have been Croatia, may have been Bosnia, who knows? The place is a little outdoor grill and the food is terrific. Eventually the bus pulls into a military type outpost somewhere near the town of Tarcin, just a spit and a holler from Sarajevo, and the driver indicates this is where you get off. Nice place, interesting place, lots of sandbags and drunk teenagers with AK-47s apparently returning from leave. Either that or the entire Bosnian army is composed of teenage drunks, which is a possibility.

There is one reasonably sober police officer trying to bring order out of the chaos. Between my German and Miller's Cohiba cigars we manage to elicit his cooperation. It seems that all the paved roads into Sarajevo have been cut by the Serbs. The only route into town is through the mountains on dirt roads, by the occasional crazy driver in his own car. Not a very promising scenario. We go over and sit at a little kiosk, that amazingly sells Coca-Cola (those people are everywhere) for four dollars a can. After a couple of hours, the cop comes looking for us.

A young kid in his early twenties with a Volkswagen Jetta is getting ready to try the back roads and would welcome the company. Of course he speaks no English or any other language we understand, but any port in a storm. We pile into his Jetta and just as we start to move out another passenger comes running after the car waving us down. He turns out to be a Godsend. He is, of all things, a psychiatrist from Sarajevo who has been attending a conference in France and is desperate to get home. Best thing though, fluent English. Now, well equipped with a Volkswagen, a driver and a translator, we start the trek through the backroads and trails of the Bjelašnica Mountains just as the sun begins to set.

If they could reproduce that ride at Six Flags it would be a phenomenon. Tiny dirt track, just wide enough for a single vehicle and quite possibly mined. Every once in a while we see headlights in front of us and our driver literally goes right up the side of the hill and stops in the underbrush while a convoy of white-painted UNPROFOR armored vehicles speeds by below. We then get back on the road, a little more relaxed, figuring if there had been any mines the convoy would have set them off. Unfortunately, soon we turn off onto a different trail, one that is too small for UNPROFOR vehicles, and our feet unconsciously start pulling up from the floor of the Jetta.

We make it. Sort of. At least we get out of the mountains and into the suburb of Ilidža on the wrong side of the airport from downtown Sarajevo. Our Jetta driver is close to home, so he dumps me, Miller, and the psychiatrist at a lone taxi that is pulled up between two earthen berms on the outskirts of town. The psychiatrist arranges for the taxi to take us to the tunnel.

Because Sarajevo is basically under siege, they have built a long tunnel under the airport to bring personnel and supplies into and out of the city. Seems like the perfect solution. The taxi driver takes us down a series of roads that have been substantially bermed on both sides to protect the traffic from small arms fire. Every house we see is liberally pockmarked with bullet holes. When we get near the tunnel entrance we have to proceed on foot, and the taxi driver goes first because he knows where the trip wires for the booby traps that protect the tunnel entrance are located, and he lifts them so we can scoot under.

The tunnel itself, what little we see of it, is a marvel. Wood floors and beams, tall enough to comfortably stand, with some underground administrative offices attached. There is a problem though. The tunnel commander tells the psychiatrist that foreigners are not allowed to use the tunnel without a written pass from the Ilidža Chief of Police. Since by this time it is about one o'clock in the morning, that isn't about to happen for another six hours or so. The psychiatrist expresses his regrets and hurries off down the tunnel. We are left in a darkened, war-damaged world of bullet-riddled houses and earthworks where there is nowhere to go, not knowing who is a friend and who is an enemy, and with just a cab driver we can't communicate with as our only local asset. At least there is a lot of soft dirt from all the recent berm building so digging a shallow grave for Miller's body shouldn't be too much of a problem.

We are saved by the taxi driver. Obviously having compassion for a couple of stranded Americans, as well as a high level of interest in hundred-dollar bills, he takes us home for the night. He wakes up his wife and she makes up the bed in a spare room where we can spend the night, and in the morning she serves fresh ground, delicious coffee before her husband takes us to the local police station to find the Chief and get our passes.

The police station is decked out in typical war zone fashion. They have leaned telephone poles against the roof to cover the blown-out front façade and to form another, more bullet resistant wall. Every other nook and cranny is filled with sandbags. We even manage to find a cop who speaks a little English, but nobody knows where the Chief is. We have a mission, and dwindling funds, and we are very anxious to get into the city. We have the cop translate to the driver that we want to go to the surface UNPROFOR checkpoint at the airport. If plan A isn't going to work, switch to plan B.

When we get to the airport checkpoint it turns out to be manned by the French. Could things get worse! The French officer in charge at least speaks some English, and he uses it to explain in a supercilious manner that no traffic is allowed into the city in the morning, that only traffic coming out is allowed until 2p.m. and then they would start letting traffic go in. Very organized. At that point it is about 8a.m. and this specimen wants us to wait there for six hours? Fuck him! So we just pick up their little red and white striped gate and start to walk. In a flurry of Gallic angst, the officer follows us onto the tarmac begging us to reconsider: "The airport is a free fire zone, there are snipers, the Serbs will shoot you." At this point we were just too tired to care. "Fuck the Serbs, everyone from the Vietnamese to the Iraqis have tried to shoot us and we are happy to give the Serbs their chance." At this point he reconsiders.

I know what he is thinking, "These two American assholes are going to get themselves shot and somehow it will be my fault." So, he has two armored vehicles come across from the Sarajevo side and escort us in our taxicab across to the city. When we leave our taxi driver in line to go back across, with 500 well-earned dollars, we try to express our gratitude the best we can with our limited communications ability. To this day, while we assume he was a Muslim Bosnian, we don't really know for sure. He could easily have been a Croat, or a Serb and we wouldn't have known the difference. That was the nature of that crazy war.

In the city, finally. First thing is to find transportation and figure out what is going on. The city has been so thoroughly shelled that every street is walled by piles of rubble, and most of the buildings are at least partially destroyed. We are picked up by a taxi driven by Fatima, a mixed blessing if there ever was one. With Fatima driving you can always be sure of one thing, the trip will be over soon. It may end by arriving at your destination in record time, or in a fiery and fatal crash. Whichever, it will be a brief trip. It isn't just that Fatima drives at breakneck speeds everywhere in the city over any kind of road, it's that she does it all while looking back over her shoulder at us and carrying on a running dialog in her 100-word English vocabulary, never looking out the windshield. At one point she looks back with a maniacal grin and says, "Fatima is more dangerous than the Chetniks, no?" We can't disagree.

We head to the Holiday Inn, one façade of which has gone in the shelling, but which is still open. We had been told that the place we really wanted to stay, the Hotel Bosnia downtown, is full and there are no rooms available. We check into the Holiday Inn, and just as we are in the process a sniper kills one woman and wounds another right out in front of the hotel. We get room keys but since we are carrying no real luggage, just small backpacks, we don't even go to the room but hurry right out the back door to the waiting Fatima for another thrill ride down to the Presidential Palace. Now we kind of understand why she is always in such a hurry.

We are met by Sanka (just like the instant decaf), the gorgeous assistant to the Minister of Defense. She takes us up to the third floor and introduces us to a deputy minister, since unfortunately the minister himself is out of the office that day. We lay out our plan and the source of our financing to the deputy with Sanka translating, and it was obvious he has a high level of interest. He wants us to come back the next day and talk to the Minister himself.

We want to see the downtown area where the mortar fire has been heaviest; we also want to see the Hotel Bosnia. Fatima has gone so we decide to walk downtown, a short but exciting walk. At every intersection is a sign warning of snipers. The Serbs tend to hide in the cemetery by the river and shoot up the cross streets, so you are only in real danger when you are crossing the street. The rest of the time the buildings protect you from the Serbs' line of fire. We are told by Sanka to be careful—lately the Serbs have made a sport out of trying to shoot people's wrist watches off their wrist but they often miss and hit the person directly. Apparently they are bored. The trick is to stroll down the block and then sprint across the intersection, stop, get your breath, and then stroll another block.

We make it to the Hotel Bosnia, bodies and wrist watches intact, and we love the place. Not only is it a hive of activity and excitement, but counter to what we have heard they have empty rooms. We decide to shine the Holiday Inn and get rooms there. I go up to the room, but the Beast of course heads straight to the bar. When I come down he is well ensconced, sharing a table with two strapping Norwegian soldiers and a rather well-known female correspondent from CNN.

The Norwegians turn out to be the UN air traffic controller contingent for Sarajevo Airport, and the CNN correspondent can drink like a fish. Miller is ordering rounds and rapidly diminishing our already puny cash reserves. I have to say though that we learn a lot about the war that night. Everybody has a story, true or rumor, who can tell?

My favorite is one about the three factions. The story goes that the Serbs were shelling the Bosnian town of Mostar but they ran out of ammunition for their artillery. The Croats, who of course were also enemies of the Serbs, had ammunition but no guns. So the Serbs rented their artillery pieces to the Croats so they could take up the shelling of Mostar. It sounded like a plot by Milo Minderbinder in *Catch-22*. True? Who knows, but it does sum up the war about as good as anything we heard. We party late into the night, there is that feeling in the air like this is the center of a world that is coming to an end, so let's enjoy it. We finally stagger off to bed in the wee small hours; we have a mission in the morning and we are ready to get on with it.

Next morning, at the Presidential Palace, we ride up the elevator with one other guy. He looks vaguely familiar but neither of us can place him. When we get off the elevator he turns left toward the president's office and we turn right to go see the minister. We realize only later that it was Richard Holbrooke, Clinton's chief negotiator in the Balkans peace process. A bad omen. The Minister is excited, not about anything we are doing there but about the fact that the president and the foreign minister are about to fly off to Dayton, Ohio. All thought of fighting the war has been pushed to the back burner. The sickly-sweet stench of peace is in the air. Talk about bad timing! Two million petro-dollars snatched from our very fingertips. No commando school, no reason for even being in this suddenly depressing war-torn pest hole anymore. As we walk back to the Hotel Bosnia we are really depressed.

We spend the rest of the day and that evening in the bar, commiserating with each other over the vagaries of fate. Fuckin' Bosnians, fuckin' Serbs, fuckin' Croats, fuckin' Bill Clinton, how dare they declare peace and smash our rice bowl. What do we do next? That is the question. About 80 percent of our cash is gone with no hope now of getting any more out of the Arabs. We sure as hell don't want to go back over to Ilidža and try to hitchhike back over the mountains. And then of course Miller has a brainstorm … ooohh shit. "The Norwegians," he cries, "they run the airport." Last night he bought them about a dozen double Cognacs each and at some point they said they would be glad to help us get on a flight out of here whenever we wanted. No better plan presents itself so what the hell, can't hurt to try.

In the morning we find a cab at the Hotel Bosnia, not Fatima unfortunately, and say we want to head for the airport. The driver takes us down to an unfamiliar part of town, mostly just piles of rubble, no people anywhere, and just drops us off. We don't see anything that looks like an airport, but he waves vaguely to our left and says, "Airport, airport." We have very little choice but to kind of take his word for it. We start walking in the direction he had waved, becoming more and more certain that he has fucked us over, when we are startled by a short blast of a car horn. Over to our left, kind of tucked up between two huge piles of rubble, is a military ambulance. We walk over and it turns out to be a Brit.

"What are you two doing out here?" he asks.

"A cab dropped us off, said the airport was nearby," we reply.

The Brit laughs, "Whoa did you get fucked," confirming our suspicions. "We're headed to the airport, why don't you just jump in back?"

We gladly take him up on his offer, climbing in the back and meeting a couple of Brit medics. They tell us that even though peace talks are starting and there is an uneasy ceasefire, the Serbs are not to be trusted and we have wandered into their territory. No wonder the taxi driver hadn't wanted to take us any further!

After about a half hour the Brit sergeant comes around back. He tells us there has been a change of plans and he is no longer going to the airport, but a Danish convoy is coming by and we can hop a ride with them. It turns out to be a convoy of four M-113 tracked armored personnel carriers, almost certainly US Army surplus. The big top hatches are open so we just jump up into one and head down the road. About two blocks later, down the same road that we had been walking on, we encounter a large group of young soldiers in dark, almost black looking fatigues with AKs and they almost all give us the finger.

"What's the story with these assholes," the Beast asks the Danish track commander, "why are they all flipping us off?"

"Oh, those are Serbian troops," he replies, "they don't really like us much but at least they have stopped shooting, for now."

I figure if we hadn't seen the ambulance, we would have really walked right into it in about five more minutes.

The hung-over Norwegians look about as glad to see us as an old girlfriend at their wedding, but a promise to a fellow fighting man is a bond of honor so they set about looking for a flight out. There is a French C-130 on the tarmac getting ready to leave for Rome. One of the Norwegians leads us through this twisting, rabbit-warren maze of sandbags that constitutes the Sarajevo airport at the time to the departure area where the French battalion commander is waiting. Nice enough guy, he tells us it's alright with him if we straphang on his flight as long as it is all right with the Norwegians. We repeat the maze, back through the airport the opposite way until we and the Frenchman find the Norwegians again. Then we go through one of these, "Well it's alright with me if it is alright with you" circular conversations until Miller breaks in and took control: "Well it seems it's alright with everybody, let's get moving." But we have one more small hurdle to cross. When we go to the departure area the French battalion commander asks us, "Where are your helmets and flak vests?"

Helmets and flak vests? "We don't need no stinking' helmets and flak vests." We hitchhiked here, walked all over town, ran through Sniper Alley, got abandoned in the Serb part of town, and now to walk 15 yards from the terminal to the plane we need helmets and flak vests?

Well we don't got 'em.

This creates much confusion and emoting among the French officers, "Everybody has to have a helmet and flak vest, how can they be here without a helmet and flak vest, who took their helmets and flak vests, there must be an explanation."

While this is going on, a couple of French NCOs find us a couple of old UNPROFOR baby blue helmets and the outside covers of two flak vests. Nobody will check to see if the plates are in there, they assure us. They are right. Outfitted in our blue helmets and faux flak vests we wander out onto the tarmac just in time to watch a Russian military jet of some kind touch down.

This event is greeted with applause and back slapping by the French so of course we have to ask, "What's the big deal about having a Russian plane here?" They tell us that with a Russian plane on the tarmac there is no chance the Serbs will mortar the runway when we are trying to take off. That makes us happy too. Another half-hour of sitting around and they finally start firing the C-130 up. We climb on board and in no time at all we are wheels up, headed for the Eternal City. No commando school, no two million dollars, but what the hell … we got this story.

CHAPTER FIFTEEN

The Spy Who Slid in from the Cold

We are fresh back from Bosnia, with a new appreciation for the absurd and bizarre. That is the one thing about our travels, one can see the world from an entirely new slant. The Beast has gone back to his lair, and I am trying to find some decent work before the meager money runs out. It is said that God looks out for fools and drunks, so I am covered on both counts, I hope.

A few days later, I am presented with a new possibility. A client wishes to sell his wares to the law enforcement community in Europe and wants to use our contacts in Europe to get a toehold. I artfully desist from telling him that my contacts are 10 years old and probably retired by now, but it doesn't hurt to try and ring them up. The first call I make is to members of the *Sondern Einsatz Kommand* or SEK in Berlin. These had been the premier anti-terrorist group with the Berlin police in the seventies and eighties, with whom I had a close relationship, sometimes too close. A finer group of suave and cultured thugs are not to be found anywhere else on the planet. I call Jochen and he is delighted to hear from me. That's always a bad omen with this crowd, but I am planning to take Miller with me so if they need a human sacrifice I am covered.

Way back in the old days, I had a couple of Harleys which were guaranteed chick magnets. The lads would be constantly coming over with requests for one or both of the scooters for recreational use. They would switch out my US plates for German and be gone for a couple of days, to return with glorious stories of exhaustive lust. It was a fair trade since they did us numerous favors and hid a few of our indiscretions occasionally. They also were playfully unconscious when it came to pranks.

At one point in the distant past I was sitting on the Kurfurstendamn, where over wine I was trying to seduce a stewardess from Swissair. I was doing fine, the gold was within my grasp, when with the squealing of tires and loud sirens, the boys showed up with submachine guns and kit, grabbed me, handcuffed me, threw a hood over my head and bundled me into one of their death vans, then sped off. This was during the heyday of the Red Brigade terrorists, so as they were bundling me into the van, they were telling the crowd not to worry, that I was a wanted figure from the terrorist group.

They are having their chuckle as they take the handcuffs off me and remove the hood. They had been having a barbecue at the police range with the latest porcine victim of their culling program of the Russian boars that roam the Grunewald forest in the center of the city. These pigs are the size of Volkswagens and occasionally consume the odd elderly couple out for an afternoon stroll, so they shoot off the offenders as a public service.

They had been making a beer run to the Kindl brewery where they had a standing tab, I am sure obtained through blackmail, when they spied me sitting with my date playing European bon vivant. I never see my date again, and that locale is ruined because the owners now think I am with Baader-Meinhof and crew. The kind of women attracted to that crowd are not appealing to say the least so no sense in proliferating that lie.

Jochen says that he will gather up the gang and we will have a warm welcome to the local market, additionally he will open channels with other groups for us. The happy news is that he now has his own Harley but he goes cold when I ask him if I can use it while I am in town, ingrate!

Now is the time for us to organize the trip. The client has a budget that is way above what I would have asked. When he mentions it, I choke and he says, "If that isn't enough, we will provide you with an American Express card, but you will have to keep receipts." At that point I feel like a pyromaniac that someone just handed a five-gallon can of gasoline and a Zippo. I can't believe my luck. I leave their offices with the card, cash, and an itinerary that will take me and two others on a round trip through nostalgia, with the added benefit of possibly selling some kit with a commission.

The first thing I do is track down Miller. He is at his usual haunt, on the beach at Laguna. Sitting at the Las Brisas, sunning himself and drinking Mai-Tais. I drop my car off and make my way to the veranda. I turn the corner, my face effused with good news, and run right into a minefield. There sitting with him is Penguini. Zounds! There is no way to avoid him finding out I have money. It is written all over my face. I try to collapse my features into an expression of destitution and despair, but it won't work. He takes one look at me and his eyes squinch, he leans forward and grasps the Beast's forearm to steady himself for the schmooze. He won't need it. Both Miller and I have no power of resistance to the Penguini. Even when your brain is screaming that his involvement will border on the insane, we can't help ourselves.

Once in a fit of largesse, when we had stopped at some Indian trading post, we bought him all this gaudy jewelry made from South Dakota gold. He is still wearing it as a good luck charm. It is so garish that he looks like Captain Kidd's treasurer, and it's working for him. I can only hope to keep the credit card with its 50-grand cap from him. I might be able to distract him with cash, send him on a separate flight via Istanbul, and hope the Turks remember him from his youth and keep him as a breeding stud at Boy's Town.

I don't even sit down when he looks at me and says, "What ya got?" I swear he can smell opportunity and cash like some truffle-hunting swine. I try to look bland and blurt, "I got nothing, how are you fellers doing?" trying to be southern as a ruse. It doesn't work. He looks at me and takes on that injured look, like you're all nine years old and you aren't going to let him play with your new bike. "I know you got something because you have that look." I want to dash my face in kerosene if necessary to get rid of "that look" but it's too late.

I sit with the two of them trying to make it look like there is only just barely enough for two people to do this job, and on a starvation budget. It doesn't matter. He launches into his story of destitution and needing work, and, and, and. After about 20 minutes of this I outline what has to be done hoping that he will see it as too herculean a task, but he uses it to justify why we *need* him.

He gets up and says he has to use the facilities, but cautions us that we better be here when he gets back. Sure Penguini, no problem, we are your pals. Not like the rest of humanity that avoids you like the bubonic plague. When he leaves I ask Miller, "So why is he here?" He looks at me with that cross-eyed look he gets when he thinks I am avoiding the obvious. "He is here, because he is our friend. You should have called ahead you nitwit."

This is true, but it is too late now. I ask Miller where the Penguini appeared from because none of us have seen him for a couple of months, unconsciously hoping cannibals had gotten him.

"He's been down in Central America doing missionary work, building a hospital or something."

"Yeah, why is he here, did he finish the hospital?"

"No, he had to leave the country."

"Why, did they discover he is the anti-Christ?"

"No, worse, he got caught faking a marriage in a Catholic country. He was just telling me the entire story when you showed up and saved me."

The Penguini comes back and we make plans for getting our travel arrangements done so we can take off on Monday. I go to the john and fish out a grand each for them and go back and give it to them, stopping the inevitable conversation that Penguini will need a few hundred extra for something, with the caution that that's their first three days' per diem and that doesn't mean lottery tickets are an allowable expense.

As we go out to the parking lot I give the Beast the credit card and tell him to make our travel arrangements, and that I will get back to him for the details and that we should be out of here Monday. We won't have to worry about the Penguini not showing up. He won't let us out of his sight lest we forget to take him. It's like he is psychic.

Our trip back to Berlin and points beyond begins in a cloud of nostalgia. Both the Penguini and I had been there with the Detachment in the seventies and early eighties. It was one of my favorite cities. We had been there at the height of the Cold War, when it was in the middle of the Soviet zone, with strictly controlled access. East Germany was a cold drab place while the West was a bright and bustling modern city. Now we are headed back, and I am excited to see how much it has changed. We each have a window or aisle seat, partially for comfort and partially to be responsible for any misbehavior.

We aren't airborne more than a few hours when the stewardess comes up and asks if the gentleman in 12A is with us. Jeff is sitting across from me so we can talk. We look at each other both considering "No habla English" but admit that he is and ask why. She says that he is making the other passengers nervous. I know she is eyeballing the empty seat next to us as a place to park 12A. She wanders off and figures out how to fix the problem. I take a huge dose of anti-histamines and sleep all the way to Frankfurt. When I awake he isn't in the seat so maybe she poisoned him and stuffed his body in the john.

We take a connecting flight to Berlin. When we land it's a dreary winter day, but we are back where we spent six years of our lives in the seventies. Berlin and Munich are the heart of Germany for me. This place always invigorates me. We get a nice hotel on the Kurfurstendamn and make ready to meet Jochen and crew the next day. The Penguini is also going through the nostalgia trip and during dinner is regaling the Beast with tales from the dark side of our earlier tour there.

He is relating what happened one night in the U-Bahn station in the Turkish sector, when a Muslim gentleman pulled a knife on him and Kato. He goes into great detail in the martial arts move both demonstrated and how they had left the body in one of the stalls in the men's room. I notice that a couple of tables actually move to another location, and am wondering if the group is thinking of calling the Polizei, that we might want to call Jochen early.

* * *

Our meetings are a success; our client's materials are needed by the law enforcement community. We host a couple of dinners for the brass, get some new leads for expanding the effort, and end up finishing early. We have a free weekend, so we decide to visit some of our old haunts. The unit has changed locations, so we drop by to see some of our former comrades. It's like old home week but they refuse to meet us at their new digs until we have met and they can threaten us with retaliation if we expose their crimes from an earlier era.

We meet at our old hangout, which is owned by a British woman who has had it for over 20 years. I swear she is still wearing the same outfit she wore when we had my going away party 16 years ago. It's all right though because she is entertaining.

The night resounds with jubilation and comradeship, followed by trying to get a cab back to the hotel whilst convincing the driver that we are not that drunk. Just like the old days.

The next day, we decide to go over to East Berlin. As military we could never go there legally, but as civilians it's a snap. We take the S-Bahn over to Friedrichstrasse, get our papers stamped and have to exchange Deutschmarks for East German Ostmarks. We are soon to find out that these have little worth outside of the government-run stores and venues. Everyone else wants dollars or Deutschmarks, if they think you are a source. We spend the day sightseeing, visiting the various places that Penguini and I had seen surreptitiously, many years before. We ask a cab driver where we can spend our trove of Ostmarks and he tells us at the Russian-owned hotels. Seems the Russians are still clinging to the fantasy that this East German paper has value. We go to one of the Intourist hotels and get properly sloshed, spending most of our East German cash. On the way out we notice a souvenir stand which features a variety of postcards with various scenes of East Berlin. They also sell local stamps. We buy a postcard and a stamp and mail from the East back to the US. All it says is, "Don't pay the ransom we have escaped," and should give the crowd in Orange County a rise.

The woman at the stand in the hotel where we purchased the cards is a very severe-looking copy of the villainess in a Bond film. She is wearing a dull brown outfit and looks like she ought to work for SMERSH. She keeps giving the Penguini the eye of suppressed communist lust. I refuse to visualize that union, so he is safe from us selling him outright.

By 10p.m. we are lit as a Christmas tree, and bundle into a taxi. It's a Trabant. For those of you who aren't familiar with this automotive breakthrough, it is produced in East Germany based on a Fiat design, and has a completely plastic body, with an engine that can barely power it. We pile in and the first thing the driver says is that he won't take anything but dollars. That's fine with us even if it is expressly verboten.

We are nearing the Soviet side of Checkpoint Charlie, when Miller suddenly says stop the car I need to get out, we want to cross here. The driver doesn't slow down so the Beast merely steps out of the car and starts to slide down the street upright and like he is the star in the ice follies, his London Fog spy overcoat flapping in the breeze like a cloak. He does it all in one fluid motion. I am amazed and terrified at the same time. I start to say something to the Penguini but notice that he too has apparently bailed. I am alone with the driver who is petrified because Miller finally falls and slides down the snow-covered street to end up against the tracks of a half-track parked next to the entrance to the checkpoint.

There are Russian and TRAPO police clustered at the terminus of his trajectory. The driver tries to pull into a side street, but it is too late, another halftrack blocks that route and we are soon stopped and I get out of the car. The Beast has picked himself up with the help of a couple of Russians. He is obviously drunker than an entire shipload of sailors on shore leave, which the Russians find amusing. The TRAPOs stand there in sullen silence fingering their weapons. The Russians are in charge thankfully. As I draw up to the Beast and his escort, I am explaining to them that he is drunk, and we are very sorry to intrude. Soon a major comes up and talks to the guards, then in perfect English says to me,

"Your friend is very drunk, what was he trying to do?" He is chuckling when he says it, so I tell him, "He wanted to cross back over into West Berlin from this side, like in the le Carré novel, *The Spy Who Came in From the Cold*."

He chuckles again and says, "That location is in Potsdam. He and you are lucky that he did it here, because they would have shot him there." He has us moved inside the office where we are obviously detained. The taxi driver isn't as lucky, they are tearing the Trabant apart and soon discover he has dollars on him, and since he is German the TRAPOs take over.

The major asks if we want some tea, apologizing that they don't have coffee. He adds that he could offer us vodka for the cold, but my friend appears to have had enough. The Beast is swaying in his seat and blissfully unaware of the trouble he has landed us in.

The Penguini is nowhere to be seen. He vanished into the shadows. There is an S-Bahn station a few blocks from here. I am hoping that he made it and is on his way west. At least he can alert someone that we are in custody. The KGB major is affable, and we are discussing what he should do about us. He is alluding to the paperwork, the diplomatic jumble, the complications, etc. I suggest that perhaps we could come to an arrangement, and he tells the other two enlisted to leave the room. He asks me what I have in mind; I suggest that a contribution to the KGB benevolent society would seem reasonable.

He smiles and walks over to the window and pulls the shade. He artfully explains that money could be construed as a bribe and that is not permitted in the workers' paradise but suggests that certain contraband could be seized without incurring complications and charges. I ask him what kind of contraband and he looks me dead in the eye and says six half-gallons of Jim Beam and 10 cartons of Marlboros would probably cover it.

Obviously, I don't have that on me. I suggest that I could probably cross over and obtain the goods, but I would have to come back over at Friedrichstrasse. If the American MPs are involved they will stop me. He says he will arrange for me to leave and then reenter with no problems. He adds that the Beast will have to stay here until I return, since he can't in good conscience release him in the shape he is in. Good: let the heathen sober up in his fantasy. I hope he wakes up and realizes we aren't with him, it would serve him right.

I am given a ride to Friedrichstrasse, and as soon as I can get to a phone on the other side I call the lads. They are still partying at the bar. I arrange for two of them to bring me the loot and I will cross over and retrieve Miller from fantasyland. I also tell them that the Penguini was missing. I am told that he had called in and had already informed them of the situation. They ought to someday do a reality show called Survival and use him as the example. I wait at the S-Bahn station two up from the crossing, and in about an hour the loot arrives. It's all tastefully packed in a cheap suitcase. I take it and board the train, cursing Miller. I realize that the Penguini's Houdini act probably saved me a couple of hours. He had done what I suspected. He had made his way to the nearest S-Bahn, crossed over at the right spot, caught a cab and alerted the lads. So he is off the hook

I pass through the checkpoint, so easy that I should have had a sign that said, Do Not Delay. I walk outside and the vehicle which had dropped me off is just pulling up. I had the guys add a couple of extra bottles and a carton to the loot, to cover anyone not in the original ransom.

When we get to the checkpoint, I am ushered inside like I am related to Brezhnev. The major doesn't even look at the loot longer than to see that there is more than what we agreed to. I look at him and merely say, "It's a tip." He smiles and says that we are free to go and that he will have the vehicle transport us to the checkpoint, adding that he wished he could let the Beast have his moment, but it would cause problems with the MPs. I agree. I ask where Miller is and he says he will take me to him, stating that he is entertaining the troops.

Across from his office is the break room for the guards. As we near I can hear raucous laughter and singing. As we get closer, I recognize who is singing. It's the Beast, leading a chorus of Buddy Holly. I open the door and he stops in mid-lyric and I tell him it's time to go, American bandstand time is over. He is given a hero's sendoff by the troops. The major escorts us to the car, thanks me for the entertainment and we are soon dropped off at the S-Bahn. I notice the driver palms one of the bottles to the commander and we are ushered through. We spend the next few hours reliving the experience with our clique, joined by members of the SEK who were on the jungle telegraph. The Beast finally gives up the conscious state and goes to sleep. I want to leave him in a snowbank.

* * *

The next day dawns with bright sunny skies. It is still frigid, but the sun warms the land and our spirits. The Penguini is delighted because Miller's stab at greatness has taken him out of the barrel. He had acted logically and adroitly, his escape and evasion were textbook. Not surprising since he is the quintessential creature of the shadows. I am having a ball telling the Beast that it cost us a case of whiskey at premium prices and it is coming out of his per diem money. He isn't buying it, bringing up historical faux pas on our part where he saved our bacon in the past, and completely ignoring the possible international exposure, criminal charges, and other cost plus items.

Our business is finished here. We have accomplished what the client wanted, made the connections that they needed, it is time to hand it over to the marketing department, with the reasonable assumption that they will screw something up and we will have additional work cleaning up their mess. I call the client and tell them that we are hosting a dinner for the police and those that helped us.

So far we have been conservative in our use of their magic card. I figure a first-class blowout will be acceptable. The decision is made to hold this gala affair at one of the old establishments that we had used back in the day as a safe house. The establishment is nestled between the residences of the British commandant and the British provost marshal. It is a palatial home that was confiscated property of the Nazi party after the war. It was the premier bordello in West Berlin at that time. Its sprawling grounds hold the boudoir, with a bar and restaurant on the bottom floor, all tastefully furnished and secure from prying eyes. The clientele act like Baptists at a Hooters, no see, no tell.

Additionally, the madame knows and likes us, especially Penguini. I swear we could make a fortune if we just rendered him down and made aftershave out of him. He attracts women better than pheromones. We arrive with great fanfare, a party of seven counting two of Berlin's finest.

The women that ply their trade are nominally from whatever refugee population is floating in Europe at that time. In this case they are Yugoslavian, Russian, or Czech. The establishment charges the client, the house takes a cut, the girl gets her cut and a portion is set aside as a retirement fund, if and when they decide to move on. It is all calibrated, taxed and health care is provided. The girls are exquisite, and they do a safe, profitable business for all concerned. We enter through a luxurious foyer, and head into the bar to await our reservation at the restaurant.

When you enter the bar there is a glass wall behind the bar stretching some 30 feet that appears to be an aquarium. The girls are all in evening gowns and circulate through as a mobile shopping display. We are seated and order our drinks. The madame comes over and asks how we intend to pay; if by credit card they will have plates which print in everything from car repair or rental, to insurance, and will separate the bill into whatever category you desire. We fish out the magic card and tell her to put the bill on at least four categories, and a dinner party with drinks for 20.

Once that is settled the night begins. Miller has never been to this place before and he is transfixed at the wares and accommodations. There is a sauna spa that connects to a pool. He is watching as a pair of male legs with a female body entwined walks across the glass wall, fully visible in what appeared to be the aquarium. He finally gets it, that if you go in the pool you are part of the entertainment in the bar.

I am totally relaxed. I don't have any desires other than to unwind from the last week's events and opt for conversation and cognac. I take off for the spa for a body massage and the pool. While I am relaxing in the pool one of the girls comes out with a hashish pipe and a bottle of champagne, compliments of the house. Why not? It's been a long week. Miller joins me after coming out of the spa in a towel with another towel wrapped around his head like a woman. He looks like a really ugly broad with a mustache.

I ask him through the warm fuzzy feeling I have about our guests. He says they are happily engaged upstairs. I ask him if he has seen the Penguini. He replies that the last time he saw him he was sitting with the madame and they were laughing their heads off at something. Why not? She has known him and me for years and she likes him. I hope she entices his ass upstairs then chains him with a leather mask and a gag and whips him like a quarter horse around her apartment. I've never been in there, but I have heard rumors. I hope they are true. It would be justice for abandoning the good ship Trabant right before the detainment.

Dawn is breaking when we finally leave and get back to the hotel. We nap, clean up and have a police escort to the airport. We spent just north of eight grand last night, all properly documented as car rentals, hotels, and dinner. I consider it a fair bonus for what we have achieved. The company will garnish over a million dollars in sales and services, probably more when it is all done. My friends have had a good time and we look like heroes. Life is good.

When we return everything goes smoothly except the CFO smells fish in the expenditures and wants to verify the receipts. No problem since the phone numbers on the questionable ones all ring into the house's switchboard by code and are answered as such. I was hired by the owner not his bean counter, and I tell him so.

CHAPTER SIXTEEN

Lah'Tay on the Spanish Main

Our fates seemed entwined, the three of us. We band together for tasks that would seem to the sane and stable as arcane and possessed. It isn't that we drift towards these as an escape; it is as if we are called to attempt what the timid would leave untouched because it seems impossible. Of course, it could be simply that we are too lazy to get an honest job. Once you have achieved what others fail at, it's hard to convince yourself that you are mortal.

We have been doing a new venue, which seems to satisfy our need to impart on others a sense of that personal satisfaction and worth. We have formed a training organization that employs us in those pursuits, and is rewarding financially. It's a great idea, but alas our albatross is the lack of being able to fund it properly to make it grow. Part of that is we are split between training professionals, and crafting programs that appeal to those that want a taste of the evil queen, Adrenalin.

We had backed into this after the Beast found an amazing opportunity with the guy we had used for our driving courses. Gordo is a former sprint car racer, who is a natural teacher. He has one of those radio personality voices and his methodology is both professional and concise. He has faults like we all do. His upbringing included growing up in a carnival. Subsequently, he has all the charm and mendacity of a shill. The boy lives, eats, and breathes racing. Every spare penny he accumulates goes into his racing hobby. You can almost smell burning alcohol and burnt rubber when he is around.

They have teamed up with a Frenchman named Charles. He is the scion of some highly placed official in the French intelligence community. Small and slight, yet he exudes an air of danger and suppressed violence. His contacts are impeccable. He finds an opportunity in Indonesia to train a group, financed by the Japanese. They spend a month putting the program together and are set to launch. The Rising Sun crowd then pulls the funding out at the last minute.

Whilst reviewing the dash of reality of dealing with the Japanese, Charles blurts out that the Japanese will spend millions paying for golf in Dubai, but they will not spend a few hundred thousand on their own security. In chaos, there is always opportunity. The Beast comes up with a brilliant plan. If entertainment pays better than reality, perhaps we should focus on entertainment.

We have formed a company called Covert Ops, which trains people in skills such as weapons handling, tactics, and defensive driving. On the professional side there are people in the security world that need these skills to better survive their adversaries, and on the general public side, those that want to garnish skill sets that one cannot find in a catalog. We struggle along training a few professional classes every month or so, until the Beast brings in an outfit from Florida that has a select clientele and offers such packages as flying MiGs, or diving with sharks, all designed to give the public an adrenalin rush and a unique experience for the right price.

The owner of this enterprise is a dour yet talented packager named Sue. She and her crew find us attractive as a venue offering to the adventure junkies. We totally agree since she can offer us as a package that will keep food on the table and maintain our skills. It is a love-hate relationship. She has her ideas on how to package us and we have a natural resistance to being offered as an entertainment venue. It does bring us into contact with the public, and some of these three- or five-day fests are a hit the first year, with more and more people wanting to attend.

It is an interesting mix of people that have money and want to taste the edge. There are some rough spots however, most of all Sue's insistence on popping in when least expected or wanted. One such night is when she makes a surprise visit after hours to the house on the base at Marana where we conduct the sessions. It is fairly late in the evening and the students are tucked in for the night when she drops in with some mundane requests and stumbles upon us sitting in and watching porn on the TV in the living room. Our mechanic and one of the off-road instructors had rigged up some sort of antenna array using reflectors and a half-wave hootefritz made from washing machine parts and a microwave.

The intent had been to pick up HBO and watch a professional boxing match but we had been unable to get a proper signal. But we could get a couple of porn channels. We are comfortably sitting watching the blue movies, swilling beer and laughing as Penguini is critiquing the actors' techniques and actions. Personally I find porn boring, but the Penguini is hilarious. I hear a gasp and look up to see the fair Sue standing there aghast. She doesn't say a word but gives us the Ming the Merciless death stare, turns and storms out of the room. Our relationship is strained at that point. She hates Gordo, finds the Beast distasteful, and seems to think I am a leper.

We are soon one of the best-rated fantasy camps available for the thrill crowd. We get a lot of good coverage, from TV and variety shows, making us a good moneymaker for the marketing firm, but we are constantly on her radar as a bad decision. It's actually sad, because she is very good at marketing us. We are what we are.

We can design and present packages such as hostage rescue, where the participants are given basic skill sets to operate as a special action team. The students are a diverse lot. We have the use of a remote base with plenty of terrain to have them plan and execute a plan to rescue people from an imagined terrorist or criminal group. One such group is made up of a firm that rewarded their top performers with our fantasy camp. One of the group is a small intense man that was an add-on. He is a top-rated cardiac surgeon, seeking a break from his profession. He is attached to the group from the big firm.

Miller gives the group their operation briefing telling them that the guerillas are actively patrolling the area around the target building. This area is scrub brush and desert. What the Beast doesn't know is that I have donned a gorilla costume and am shadowing the group in the dark. Everyone is armed with paintball guns. I am occasionally firing their butts up then slipping to the next stand of vegetation. I hide in the bushes directly along their route. The group passes me in the dark, and I count them as they pass. The last one in line is the surgeon. I reach out and envelop him in a hug and he passes out from fright. I check his pulse, panicked that he has had a heart attack. His ticker is still working. I wait for him to come to. He leaps to his feet and he runs off back towards the base.

He breaks in the room where Miller is waiting, and blurts out, "There are g-g-g-g-g-oooorillas out there!" The Beast looks at him and says "I know, I said so in the briefing." The guy says again, that there are gorillas. Jeff still thinks he is saying guerillas and ignores him until he realizes the guy is serious. The Penguini repositions me in our house, and goes to tell Miller that some wild animal is in the back bedroom. Of course the Beast, ever pragmatic, comes in and begins his Wild Kingdom search pattern. I am hiding behind the door. He slams the door open, barely missing smashing my nose flat, and enters the room. I reach out and put one giant fur-covered hand on his shoulder and he leaps backward into the hall so fast that he actually dissipates. This, of course, is part of why we are distasteful to the marketing maven, but the participants love the variety.

We have a group from the *Tonight Show* that show up and have a great time and Miller gets to go on TV. He's good at this and it only serves to increase our popularity.

The Beast is determined to cash in on the idea and two former participants agree to fund a fantasy camp in the Caribbean, based on the same theme and marketed to the cruise crowd. The cruise lines are always looking for shore excursions that are theme based and appeal to men. We start to market the idea to them and they are enthusiastic about it.

He sends Charles and a man that he recommends, who is a former spook and South African infantry officer, to go to St. Thomas and set up operations. They are to find a spot where we can build our bunker system, rent a house, buy a truck to haul our stuff back and forth and make connections with the proper authorities and venues.

After a couple of weeks with no feedback he sends the Penguini down to see what's going on. Charles has disappeared but has left a strip map of the area that he has chosen for our amphibious landing. It is totally unusable. The beach is about a meter wide and ends in a vertical cliff which screams liability lawsuits. He goes to the house they had rented. It is unlocked, deserted, and the TV is still on, no sign of either of the advance party. We find out that the other guy had run up a bill at the video store for porn movies, it seems they had a nice vacation, ran through their budget, and just left. There is a dilapidated Toyota pickup in the driveway. It looks like a survivor of a demolition derby but it runs.

The Beast and I fly down and we start fixing what they had failed to do. We are busy shopping for rubber boats, engines, paintball guns, camouflage uniforms, and all the equipment we need for the fantasy.

While we are shopping for a site, we ask our real estate agent about a small island that could be perfect for us, off the east end of the main island near the town of Red Hook, where you can catch the ferry over to St. John. "Oh no," she tells us, "that island belongs to a very rich man with many young girlfriends, it would not be for rent." She adds that "President Clinton and his wife are guests there, President Clinton very much." This is 2001, and the residents of St. Thomas all knew that something was going on.

We find the perfect spot, which is a small island off the north coast of the main island, called Inner Brass Island. It had been a PT boat base during World War II and is deserted. We track down the owner who is part of the French community that inhabit the back side of the island away from the main town and tourist center.

St Thomas is a weird set-up. The predominantly Black community is in government, running a white-owned economy with the elan of a despotic upper class. The Frenchies who are white and mixed race inhabit the far side of the island and the two segregate themselves based on economics. The capital is for the tourist mobs, and has so many jewelry shops that the island is in danger of flipping over. There are historical places like Blackbeard's Castle, the garishly painted Dutch reform architecture, and the thrill of being in the middle of the Spanish Main. There is a stone bench on a vista looking out to sea called Drake's Seat, where supposedly the eminent privateer used to post a watch to espy fat galleons heading for España. The tourist shops at the top are staffed by immigrants from another island who are the biggest lot of surly racists on the planet. The locals hate them. All this in all a tropical wonderland.

All the Frenchies have a favorite watering hole perched on the precarious road that winds around the mount facing the capital. We are invited to one of their shindigs, in the hope we can romance the owner of Inner Brass to lease it to us. It helps that we have hired her brother to help with our venture. The reception to our wishes is at first cold. The band begins to play and they are doing fifties music and the ladies are swing dancing with each other. A few come over and ask if we would like to dance. I am out of that category as I dance as if no one is watching. Finally, one peels the Penguini out of his chair.

He is wearing a black Hawaiian shirt with a huge dragon in white going down the front. He glides out on the dance floor and within the first three moves establishes that he is the king of the boards. His movements are smooth and precise his twirls are moments of magic. The ladies are now following his every move. The dance ends but the one who hit the jackpot refuses to let him leave. They are queuing up to cut in on each other like a pack of rabid cats. One of them comes by and looks disdainfully at us, like the rubes we appear to be, and coos, "That man has the magic moooves." By the end of the evening he has charmed the female population of the Frenchies. We get the lease on the island for 24 months. *Fini*!

He comes back to the table soaked in sweat and begs off the next three dances, and out of gratitude we buy him a couple of Bushwhackers to rehydrate himself. It's a unique drink to the island, tastes deceptively like a chocolate shake and has enough rum and kahlua in it to deaden even a keel hauling on a stiff rope. By the end of the evening we clamber back into the Toyota followed by a retinue of hormonal women chanting his name like he is the Bhagwan of their newly formed cult, and return to our humble abode on the other side of the island.

Our next task is to make a deal with the man who owns the resort facing the small island. He is a big man with a ruddy complexion and a disconcerting resemblance to Hermann Goering. He has a nasty reputation that makes the police and the rest of the island fearful of him. He is also batshit crazy. He once got drunk and on a bet followed a cruise ship all the way to Puerto Rico on a jet ski to chase down some idiot who skipped out on his bill. He likes us though and we get along with him just fine. We are able to stage our adventure from his resort, use his barbecue and pavilion for our champagne brunch, and store our equipment there.

We have to now build the pirate camp we need for the scenario, where the captain of the cruise ship is being held for ransom. We have decided on a bunker system with connecting trenches gussied up to look piratical. In the scenario a select number of the participants will play the scruffy kidnappers and the others, dressed in camouflage and face paint, will make a raid in rubber boats, land, recon, then raid the camp, rescue the hostages, then back for brunch. We can handle 40 to 50 at a time, twice a day. There are five cruise lines that call on the island and we have signed all of them up.

First, we have to build the camp. We have nothing but hand tools to do the construction. The work is back breaking and the temperature after 9a.m. is tortuous. It's high summer, and we are scurrying to have everything done when cruise season starts in October. We soon come to the conclusion that we need help. Hiring Dominican labor is out of the question since they will work long hours but will steal you blind. Our saving grace is that we find four Australians that are temporarily short on funds and need the work.

Australians are an odd lot, but they work like demons especially if you pay them well and sweeten the pot with a ready supply of free lager at the end of the day. It is still bloody hot. To add to the misery there is a species of tree that is caustically poisonous. If you rub up against the bark you get a chemical burn that is both painful and gives you the appearance of having bathed in meat tenderizer. The trees are everywhere in the old pineapple paddocks we have chosen for the camp.

The Aussies have come up with a preventive measure that consists of smearing themselves with mud from head to toe, mixed with some plant extract. Aussies love dirt, it's part of their culture. They show up in their raggediest clothes, smeared in their ochre-colored concoction. They look like a Georgia chain gang, but they work like demons. Both the noon meal and work's end are marked by all of us staggering down to the shore and falling into the ocean. We have to curtail this after we spot reef sharks attracted to the splashing, cruising in to look over the menu. It's a long hard two weeks of back-breaking work.

Since the heat and humidity get intolerable after about noon, we knock off in midafternoon mainly because we are all dehydrated and bone tired. The soil on the island is like concrete but we manage to hack our little theater of the bizarre out of the unyielding dirt.

I keep wanting to get started earlier and earlier, like get over there by 5:30a.m. when it is reasonably cool and work like demons then quit when it gets too hot. There is one impediment. The legend of Penguini, of magic moves fame, has spread all over the island. He takes immense pleasure in stopping at the coffee bistro on our side of the island, before we make the 30-minute drive over to the resort. He is enamored with the lady that makes his ritual latte before we start over. It doesn't open until 7a.m., and he won't slurp it on the ride because he needs to relax and prepare his day with his latte tryst. We lose an hour of cool temperatures in this ritual. I want to smear him with chum and have him tow one of the boats over doing the side stroke. My protestations are to no avail and he reminds us that he will pull our arms off and use them as whips if we keep pestering him. I have daydreams of him making his famous magic, while being pursued by a dozen reef sharks.

We soon have our set fully functional. We have arranged for a tour by the various social directors on the ships. We bus them over and load them onto the boats to take them to the island. Only six show up for the first run so we are in one semi rigid.

Since the Frenchman we had sent to do the advance work had blown our budget on a soirée, we had been forced to buy on the cheap. We had gotten these boats used and had spent a couple of days patching and refurbishing valves etc. The one we have chosen has a reputation of blowing valves and deflating. This isn't bad if it is only one cell because the trip is about a half a mile to the island. I have tastefully prepared for this eventuality by hooking up a manual pump to the faulty cell so we can keep it inflated. We start for the island with the tour directors having a marvelous time with the Beast as the coxswain.

We are halfway to the island and the bloody valve blows and the core shoots in the air like a Polaris missile. Soon the middle of the boat with the cruise ship crowd is deflated and they are piling towards the back where the Beast is manning the poop deck. There isn't enough room so they stampede over me pushing me to the deck, in the direction of the where the Penguini is perched on the inflated bow cells. He has armed himself with a paddle. I am yelling at the Beast to return to the resort but he ploughs on like a modern-day Ahab.

The Penguini seems intent on holding his ground which would be bad form in the way of customer relations, so I tell him to let the crowd up where he is. He shakes his head no and points off the starboard. I follow his motion and look overboard. There just below the surface is a 12-foot tiger shark, effortlessly following our punctured raft and eyeballing the tasty treats warming up the meat. We make it to shore and unload the boat, with the Beast droning on, "Wow wasn't that exciting folks, maybe we should hire the shark for each crossing." He gets nervous laughter at his jibe and we start the tour. I use the radio to call the brother to bring over a fresh boat and by the time the cruise people come back we have our newest vessel in the fleet ready to take them back.

Our opening day is next week and we are pre-booked for two months with a hundred people a day, at 95 bucks a pop. We will be in the money next week, and out of debt the following week. The Beast has returned to the States to pick up our supply of glossy brochures which we will seed in all the hotels and resorts.

The Penguini is up and getting ready for his latte and I am having a strong coffee in the early morning, staring out to sea, and flush with the satisfaction of our venture and the beauty of the Caribbean, fully displayed with a slight breeze blowing. We have moved the TV up here under the shade and I am watching the morning news. It is September 11 and our opening is in two days. I am half aware of what's going on in the news and am absently watching. I am enjoying my coffee and have two days off before we start in earnest. I watch the first plane plough into the World Trade Center, and it really doesn't register what is happening. As I watch, the attack on our country unfolds.

This is Pearl Harbor all over again. I yell down to Penguini, to get up here fast. Something in my voice must have told him that something was terribly wrong, because he pops up with shaving cream still around his ears and a towel in his hand dabbing at the remnants. We stand and watch as the tragedy unfolds and thousands of our fellow citizens die a horrible death.

I call Miller and reach him. I tell him to turn on his TV and see what we are looking at. Now the three of us are watching and realize that we are at war again, and at the same time realize that our venture is totally screwed.

The chaos that follows is happening all over the world. There is panic and the cruise ships are calling their passengers back with their horns making a funeral dirge, competing with the hubbub. The authorities lock down the port and the airport and we go from fun and frolic to war footing. The local coast guard station is a hive of activity and people are searching each other out as if talking about it will somehow make it all go away.

We are in a rage as it becomes apparent who has attacked us and why. All of us have lived or worked in the Middle East. The stories of Muslims in our own country celebrating incites a rage in me that I haven't felt since the war. This cancer has crept into our societies and now is at our throat. We can only hope that we retaliate and retaliate so that the memory of it stays with the survivors for generations.

I think of Admiral Halsey's comment when the carriers came back to the destruction of Pearl Harbor. He softly commented to his aide that when we were done Japanese would only be spoken in a whisper. I feel the same when we are done with this scum the Adan will be spoken in a whisper and only in hidden places and this cult will only be a memory.

The Penguini, ever the pragmatist, looks at me and comments that maybe we can sell all the motors and equipment to the Frenchies, because tourism is dead for at least a year. He is right. The subsequent week proves it out. The cruise ships steam out with a destroyer escort, ferrying people to either Miami or San Juan to eventually get flights back to the shell-shocked families and friends. They then go to an anchorage in the Bahamas. A single ship arrives four days later picking up tourists stranded by events on the smaller islands; it also has a naval escort.

The Penguini and I manage to get a flight out the day the airport is declared safe for flights entering the United States. We are standing in the waiting line for customs and immigration. There is a large sign behind the officer checking our papers that says marijuana and drugs are a felony offense.

What is incongruous is that the wall behind him doesn't reach the roof beams and there is a gap of a few feet. On the other side of that wall is where the cabs gather. The whole area reeks of weed. I make a crack about it to the officer and he just gives me a sour look and waves me on my way.

Our venture is a bust. We are headed back to an uncertain future, our country has been grievously attacked and the enemy is within our borders as well. We all hope we can get in on the fight. I want to smash them into dust, destroy their cities and make them quake in fear at the mention of our country. They deserve it and their apologists even more.

We eventually have to send the Penguini back to settle our lease. We end up giving the Frenchies everything, in return for cancelling the lease. Our tomorrows are gray and dismal, and we are once again cast back into the crucible.

CHAPTER SEVENTEEN

Soaring Hopes and Soaring Temperatures

The aftershocks of 9/11 wipe out more than just our shore excursion venture in St. Thomas. That day completely changes our entire world. We also shut down the fantasy camp in Arizona, it just seems too close to reality to make good entertainment. Suddenly there is a real need for us to return to our roots, and the new conflicts, the global War on Terror, the war in Afghanistan and later the war in Iraq offer a plethora of opportunities.

Our first big try is a multi-million-dollar government contract for training the security detachments for the Iraqi Council of Judges. Of course, it is issued as a Request for Proposal by the Defense Department and put out for open bid. We assemble a team to prepare a proposal and go looking for a training site and sources for all the equipment and vehicles we would need. Jeff flies off to Amman, Jordan and Baghdad while the rest of us get down to the scutwork of proposal preparation.

In Jordan, Jeff stays at the Four Seasons Hotel in Amman where there is actually a bar in the lobby with a selection of good single-malt scotches. He interfaces with an uncle of the king who is able to fix him up with most of the necessary equipment and turns him on to what appears to be a good training site in a not too dangerous area. From there he flies on to Baghdad. At that time there is still a danger from shoulder-fired surface-to-air missiles so planes flying into the Baghdad airport don't use a natural glide path. They position themselves directly above the airport and then descend in a steep corkscrew straight toward the ground, pulling out just in time to touchdown on the runway. Exciting! Especially your first time.

At Baghdad airport he is met by a security detail we had hired for him, a driver and a guard from one of the larger private military contractors, and he is driven into the green zone. The driver is a good guy, a cop from Santa Barbara, California. The guard is a younger kid who claims to be former SF but it doesn't ring quite true. He is a little trigger happy and Jeff is constantly admonishing him for drawing down on every single person that comes within range, it gets embarrassing. Jeff spends a few days meeting with the contracting officer and getting the lay of the land before returning to the States full of optimism and ready to go. He thinks we have the contract all but in the bag. His enthusiasm infects the whole team and we all start looking at brochures from the Newport Beach Mercedes dealership, picking out the color for our SL 500s.

We prepare what we think was a really good proposal. In addition to the training mandated by the RFP we add a complete Evasive Driving and Transportation security section that we doubt any competitor could match and still manage to keep the cost reasonable. Now we wait. It takes a few weeks to get the news: WE HAVE WON! Everyone on the team is immediately put on the payroll and we start to order equipment through our contact in Jordan. We have to gear-up fast. One of our team is sent back to Virginia for some necessary special training and we are moving full speed ahead when suddenly, the rug is pulled right out from under our feet.

A British security company who had also bid the project, and who had obviously lost, had protested the award. This being our first really big government contract—we had done many small contracts but nothing like this—we weren't even sure what that meant. We talk with contracting officers in Baghdad and with a representative from the government accountability office in Washington and are told we have to re-submit a new proposal with our "best and final price." What? The price we submitted was our best and final price, we didn't have any slush in there at all. So, we just resubmit our proposal and our price a second time. We lose.

We learn later that the Brits, whose original bid had been well over a million dollars more than ours, came in with their second bid just 100,000 bucks below us. Strange when you consider the original bids were supposed to be secret. So they just barely beat us out on price, although they still didn't have any of the driving modules which are by far the most expensive module to provide, and since we had already spent a bunch of money while we thought we had the contract, the government has to pay us one point two million for our wasted efforts, which is *waaaay* more than the few bucks they saved by moving the contract. Can anybody say corruption? Can anybody say, "the fix is in"?

Almost two years of small, subsistence contracts keep the core of the team together until another large contract comes within our grasp. This time it is the US Marine Corps and it comes through our old friend from Algeria, Cosmo. The Marines are being dispatched to Iraq in droves, and the mandatory training programs at the 29 Palms Marine base in the California desert are booked solid by the infantry and armor units. The entire First Marine Logistics Brigade at Camp Pendleton need an alternative and they need it quickly.

Once again, we assemble the entire team, a bunch who had been on the Baghdad project as well as a few new ones. This contract requires about eight instructors, four pyrotechnic specialists, 20 role players, a medic and a headquarters element to train about 1,600 Marines. Our usual training site in Arizona just isn't big enough for something this size, so we look around and found a nearly abandoned iron mining complex out in the Mojave near Desert Center. It has everything: deep abandoned mine pits that could be used as ranges and an entire ghost town that could be used for urban warfare training. We make arrangements to lease the site and construct an entire proposal around three specific training venues, a live fire range using .50 caliber machine guns mounted on seven-ton trucks, a 7.62mm live fire range involving a revolutionary moving target system we invented, and an urban warfare scenario with pyrotechnic simulated IEDs and RPG shots and a mob of role players. We submit our proposal and are almost immediately granted the contract.

There are four or five houses in the ghost town that are habitable. We also rent a house at Lake Tamarisk, about 12 miles from the training site. It is enough to house our staff; the Marines will build a large tent city on a flat area adjacent to the training site. Cosmo stays back in Orange County and Miller takes charge at the training site where we go to work building ranges and preparing our quarters, all while trying to survive the summer heat in the Mojave. On a good day the temperature would only get up to about 110 degrees, our record was 127!

One of the hotter days just happens to be the day we need to tow five junked cars we had purchased at a junkyard in Desert Center up the 12-mile paved road and then another three miles across the mine site on dirt roads to where they will be set up as targets for the .50 caliber range. We have done four, and it is hot, dirty, back-breaking work. Everybody is tired and soaked with sweat, but the Beast wants to get that one last car out there that day.

In order to tow the cars somebody has to sit in the junker to steer. It is the nastiest job, since of course the junkers are not air conditioned. Miller volunteers to steer the last car since he is the one insisting we do one more. When he gets in, he finds the windows don't go down, so in typical tired Beast rage he just kicks both front windows out. Why not, we are just going to shoot them up with .50 cals anyway.

It is a good idea on the pavement: the breeze keeps the interior down to a comfortable 120 degrees. But once on the dirt roads, dust fills the inside of the car and the temperature skyrockets. Miller starts swinging from side to side, bouncing off the berms on each side of the road while those of us in the comfortable air conditioning of the tow truck are oblivious to his plight. We drive down into the mine pit and up the other side to the target area, finally coming to a stop at the dropoff point.

Penguini jumps out and walks back to the car, only to find Miller's unconscious body. Penguini pours a bottle of water on his head and he sputters back to consciousness at which point he immediately turns his head and pukes all over Penguini's pant legs and shoes. "Son of a bitch!" yells Penguini as he jumps back and slams the door of the car in Jeff's face. The rest of us are collapsed with laughter, but I finally go over and pull the Beast out of the junker. He is in bad shape, heat stroke, so we give him SF-style medical attention: we pour more water on him and then insult him until he gets up and goes back to work. The Beast won't be our last heat casualty:

on the first day the Marines arrive to start training we lose 11, one as he is getting off the bus. It's hard to express how hot it is in writing, you just had to be there.

Shortly after the majority of the hard work is done, building the ranges, building a mosque in the ghost town, laying out the grids for all the pyrotechnics, helping the Marine advance party set up their tent city and their mess hall and locating and fabricating an ammunition dump out on the backside of the site, etc. Cosmo shows up with his new "assistant." Since the financing for the program was run through Cosmo's company he naturally thinks he is in charge and needs a staff. Miller who does all the real work is too independent; Cosmo needs a lackey. In this case it is an Asian woman who had attended Cosmo's alma mater and looks like a deckhand on a Korean fishing smack. She is also the dumbest and least decisive person any of us have ever met.

The two people who get the worst of the heat are Gordo and Greg down in the Escalation of Force range. The two had fabricated the range from a vague idea of Miller's which they first said was impossible but then in the tradition of good American garage mechanics they found a way to make it work. It involves pulling a car on a sledge they designed about 300 meters toward the back of two parked Marine Corps trucks. The idea is to train the Marines on procedures for an unknown vehicle closing fast on their convoy from behind. When the car starts moving, they have to wave it off with two red paddles, if it keeps coming they fire a warning burst in front of the vehicle, if it still keeps coming they light it up. Since we only could have two Marines at a time on the guns, over the six weeks they have to pull the target car over the 300-meter course about 800 times. It works like a charm and the Marines love it, but it has one drawback. Since it is live fire we have to locate the range in the bottom of one of the abandoned mine pits in order to have large backstops on three sides to contain any poorly aimed rounds. Down there the heat is magnified and frequently gets close to 140 degrees. Gordo and Greg take turns, one hooking up the tow chains while one drives the air conditioned tow truck, but a 10-hour day is still brutal beyond belief.

Penguini runs the urban operations ranges where he never fails to impress the Marines with his imaginative placement of IEDs. He also has all the Arab role players, most of whom came from an Iraqi expat community down in San Diego. When the Marine convoys drive through the town they run out of the house screaming invective in Arabic and throwing big chunks of cork to simulate rocks. These demonstrations are generally meant to distract the Marines from a more deadly ambush by IED or RPG and small arms coming up. This set-up puts Penguini in daily contact with Cosmo's assistant, who Cosmo has put in charge of housing and feeding these role players, and needless to say it is not a comfortable relationship.

She considers all of us barbarians because we don't hold MBAs, but she is particularly insulted by Penguini who treats her like a servant. She drives around the site in a Prius that Cosmo has rented for her because the two of them think it was important to be "green." The rest of us drive big gas or diesel pickup trucks. She is always concerned that the uncouth horde of instructors will eat the food she purchases to feed the role players, so she is afraid to put their food in the refrigerator in the house. One day she buys a bunch of salad makings, heads of lettuce, tomatoes, other organic green stuff, and to "keep it safe" she puts it in the back of the Prius. Right under that severely slanted rear window those things sport. Of course, under the direct sun on a typical 120-degree day it quickly turns to compost, brown and slimy, further cementing the instructors' opinion of her mental capacities.

One of the main problems faced up at the site is that it is an abandoned iron mine. The ground everywhere is covered with shards of iron ore, usually sharp, that are absolute hell on tires. Miller drives around with the bed of his pickup full of wheels and tires and spends half his day delivering spares to the rest of the instructor staff who have had one or more flats. On one especially bad day he has given out every good tire he had and filled his truck with flats to be fixed. He is driving back down to the Lake Tamarisk house when somewhere around the halfway point, sure enough, he gets a flat. He calls up to the office to get help and gets the assistant on the phone. "I need some help out here on the main road," he tells her, "I'm about three miles from Lake Tamarisk and I got a flat tire. I gave out all my good spares and now I don't have one for myself." She of course starts to dither: "Well there is no one here but me, Cosmo is out on the site somewhere and I'm working on budget numbers and I want to finish these but maybe I could go and look for someone to help but I really should finish this first, blah, blah, blah." At first Miller is cool-headed and reasonable: "You know it's 118 degrees out here and I really don't want to stand around forever, you could just come down and pick me up."

"Oh, I'm much too busy, I barely have time to look for someone to come help you but just wait, I'll see what I can find."

Suspecting what is coming, Miller starts walking through the stifling heat toward the lake. Several other times she calls back with more dithering and blathering, gradually raising the Beast's internal temperature until it matches the temperature outside. Nearly an hour goes by before she calls and says, "Okay, I just can't find anyone so I suppose I can come down there and pick you up myself." The Beast answered, "Don't bother, I'm just walking in the front door of the house at the lake."

For the most part the training goes smoothly; there are small daily problems to be addressed which is typical for a venture of this size, but there is only one really serious incident. Two Marine Corps trucks are heading to the ammunition dump, which is out in an isolated spot on the backside of the site. Both drivers have been there before and they are navigating from a hand-drawn strip map. The trucks are speeding across the dusty roads, way too fast for the conditions. They just happen to pass Miller and he falls in behind them, wondering where they are going and why they are in such a hurry. The roads out on that part of the site are like silt, and the cloud of dust thrown up by the lead truck totally blinds the driver of the following truck. Sure enough, the lead driver suddenly sees the ammo point and slams on his brakes. The following truck doesn't have a chance and rams into him at full speed.

Both trucks are full of Marines, the lucky ones thrown clear although many are injured when they hit the ground. The driver of the following truck and a few others are still in the vehicles and seriously injured. Miller pulls up a few seconds later and immediately gets on the radio for help. We have two highly qualified medics on the instructor staff and Penguini rounds them up and we race out to the site. It is bad. The medics set up a triage area, Penguini and I go across the road and set up a landing zone because Jeff has already called in for air evacuation of a couple of the more seriously injured.

Next to arrive are some officers from the training office along with the commander of the battalion that is training that day, a lieutenant colonel who is apparently some kind of up-and-comer in the Corps. The Marines have corpsmen attached, but they have yet to arrive by the time Penguini and I are landing the first medevac chopper from 29 Palms. Apparently, the radio transmission about a vehicle accident and air evacuation had been patched through to the CHP, and the next to arrive is a highway patrolman in his cruiser. The Lieutenant Colonel immediately goes over to the cop and tells him, "That captain over there is in charge here, not me, if you need any information talk to him." We are amazed and Jeff mutters under his breath, "Nice leadership." The Lieutenant Colonel must have heard him because he gives him a withering look.

We get the two worst cases on the first bird and it takes off, but we realize we will need a second bird and call it in. Still no corpsmen and our two medics are way too busy with the critical cases to see all the more minor injuries. Miller is impatiently pacing and looking at his watch when the Lieutenant Colonel comes over and says, "No use looking at your watch, that isn't helping anybody." The Beast replies, "Well do you know where the hell your damn corpsmen are, they should have been here by now if they were walking." At this of course the Lieutenant Colonel goes into full leatherneck officer mode and starts into a full-blown ass chewing, starting with "You can't talk to me that way." The Beast just looks back at him and replies, "Don't try your parade-ground shit on me, I'm not one of your people and I've had my ass chewed by four-star generals so I'm not impressed." The Lieutenant Colonel is shocked: he stares at Miller for a few seconds, turns on his heel and huffs off. Eventually the corpsmen show up, the critical are air evacuated and the walking wounded are ground evacuated, and things gradually return to normal. The captain comes over to Miller to ask what he had done to piss off the Lieutenant Colonel so badly. Jeff tells him and it is obvious he is quite pleased. Seems the more junior officers don't like the guy much anyway.

The days grind on. At the halfway point there is a weekend break and many of us use the opportunity to head over to Palm Springs to relax for a couple of days. Lying around the pool, dinner at Flemings, plenty of alcohol and then back out for three more weeks in the crucible. The heat and the long days are taking a toll on everyone. Every day the whole group is surlier, more short tempered and more willing to show it. Cosmo and his assistant start to become more and more the focal point of everyone's ire. He has given himself the callsign of Big Dog on our radio net, which of course is turned into Big Hog when he isn't listening or sometimes even when he is. As a result he focuses all of his animus at Miller, who he thinks is encouraging the others in their contempt. But despite the fractures in the cohesion of the team the training proceeds smoothly, and the Marines are happy with the program. So much so that they produce a contract for a second iteration the same size as the first just a month later.

We all go home, expecting to be back in a month doing it all again. It should be easier the second time because we worked out all of the bugs on the first go around and because the hottest part of the summer is behind us. Then Cosmo tips over the apple cart out of pure greed and spite. Jeff goes to the office the morning after he arrives back in Orange County and finds no one there, and the lock on the front door changed. When he returns to his apartment in Newport Beach, he finds the contents of his desk in a cardboard box on the lawn. Cosmo won't return his calls, but he contacts the rest of us. He actually thought the whole team would come back and work for him on the second program, even after the rude and underhanded way he had treated Jeff. Well, to a man we all tell him to stick it up his ass, we were only barely tolerating him because of Jeff in the first place.

Cosmo is left with a contract and no team, no knowledge, and no way of fulfilling the obligation. He thought the team members would want the money so much that they wouldn't care what had happened to Miller, but he was wrong. Now he has to hustle to put together some kind of rag-tag group of the inexperienced and the desperate. Nobody in the Special Operations community would touch him with a 10-foot pole so he finds a couple of retired Marines, a couple of wannabes, and whoever else he can dig up at the last minute, and tries to teach them the systems that had been designed by experts. He even tries to patent the Escalation of Force range designed and built by Gordo and Greg. What a slimeball.

None of it works. The Marines are aghast when they show up and this new team is so much less qualified in all of the necessary disciplines. About a year later Miller attends the retirement ceremony of a chief warrant officer down at Camp Pendleton who had been part of the Marines' training management staff. It happens that the colonel who commanded the entire brigade is there along with a captain who had replaced the captain who had been on site at the truck accident. The colonel introduces Jeff to the new captain: "This is Jeff Miller, he ran the good Eagle Mountain training." As far as I know nobody on the team has ever talked to Cosmo again.

CHAPTER EIGHTEEN

Port au Prince is Dying

There are times in your life where you are in the place where you can make a difference. You look at the odds of both success and the possibility that you will perish because of your decision. It's not a toss of the dice, but rather what your core values are that determines your actions. The ancients say that the Gods favor the bold and the crazy, so we should be covered.

We are working for a large resource developer in the oil and gas business, with whom I have had a relationship for a number of years. They are a good client, and I have always kept my relationship in line with specific tasks, rather than milk them for a retainer over an extended time period. That's best left to the beltway crowds, who have the ethics of a chicken thief. Our reputation is for quality services, performed for a reasonable price, then we are available when they need us again.

When the wars started, we had begun to provide goods and services that would be needed by the force that was sallying forth to confront the enemy that had attacked us. We were in a unique position since we had the resources to provide. One of the key items that we were able to offer was armored vehicles to the contract force that was growing around the destruction of the Iraqi regime and the so-called nation building touted out of Washington. We had a fair price and a delivery schedule that allowed us to get a small piece of the contracts being let.

I had personally gone to the war zone to ensure that we delivered and provided per our contracts. Despite being in the cheering section for our government's action, I saw first-hand the greed and corruption that had grown around those contracts. When Bremer landed as the de facto governor of Iraq he brought 300 tons or so of hundred-dollar bills to fuel the rebuild. This was outside the DOD and DOS efforts and funding. The crowd of thieves around this money was a free for all that stretched from Congress to Baghdad and involved the wholesale theft of the public coffers. There were a lot of good people that went, served, and sometimes died in this effort but there were an equal number with their hands in the till up to their elbows. We lost friends and dear comrades while the thieves got rich. In the end

I was ashamed of those countrymen for their deceit and avarice, for they were traitors to what we stand for as a Republic.

We had helped develop a technology brought to me by a friend that had been in both CCN and the Phoenix project in Vietnam. A true gem of a man and one that you would want at your back, with cold steel in his hand, when it gets ugly. We had sent the technology downrange with Special Forces to test and were confident that it performed beyond its capability profile.

We now had a situation where the client was suffering armed attacks against its employees and had already lost personnel in the attacks. The military and the police were at a loss to stop the attacks as the area was rugged and covered with tropical jungle, thick impassable jungle, that ate men and equipment. After a review of the situation our first suggestion was to put tracker teams in to hunt the perpetrators down, kill them, take prisoners, and aggressively interrogate them until they led us up the food chain to whoever was behind the attacks. This of course, caused the legal department to have apoplexy since it was direct and effective, not in line with their normal methodology of wasting funds while talking endlessly about the problem.

We had introduced the idea of using the technology since it could find and track humans out to a kilometer with no false positives. It wouldn't alarm on animals, just live humans. They had agreed to test the technology since it was a passive approach, then let the local security forces do the dirty work.

We are in the mountains east of Tucson, Arizona testing the devices using the Penguini as the target in rough terrain with good undergrowth. We want to deploy the devices with an aerial platform so we are driving on mountain roads aiming the device out the door of a van at the valley below, and picking up the targets then following them. Once they get tagged, they step out to be visible. We have had a few glitches and are working out the details. On the third day our "manager" calls us about discussing another mission for the devices.

Our manager is a six-foot-eight-inch Hessian, who after leaving the German Army had attended one of America's premier graduate schools. In fact, it was the same alma mater as Cosmo, the Weasel. He is the total opposite of Cosmo though, dedicated to his employer and profession. We formed an immediate bond. I believe that he was always wary that we were loose cannons but had enough self-confidence that should we get out of line he could handle the situation by killing us all and eating the evidence. In return we go to great lengths to protect him from our foibles and indiscretions.

He informs us that there has been an earthquake in Haiti, and the company wants to know if we could deploy the technology quickly to the disaster area. We are confident that our group can handle it, maybe with a couple of add-ons with special skills. There is just one obstacle: our manager's boss is a former West Point ringknocker, and displays all the reasons why that finishing school produces a few brilliant officers and a host of mediocre sycophants. He is a typical example of thinking only inside the box. In his case the box is double taped from the outside.

We quickly assemble a team which include the Beast, Penguini, Magic Mike, another SF alumnus and former Agency contractor, and a diminutive veteran of the Afghan wars, who is a dead ringer for General Billy Sherman and mad as a March hare, named Gator.

Our Hessian proctor explains that the company is going to provide us as an adjunct to the rescue effort for both good will and good publicity. I don't question people's motives but the plan makes perfect sense. We have used the device before to locate people in containers, in tunnels, and in rubble, so we are confident that we can help save lives. We have a lot to do before we can launch, but we organize, split tasks and are soon on our way to Miami from where we will launch.

Once on the ground our "manager" whips out the company plastic and we set out to get the equipment we will need as a rescue crew. We buy REI, lamps, sleeping gear, water purifiers, dried rations, camp stoves, hand tools, ropes, snap links, every imaginable device that will assist us in getting into the rubble. When you need this sort of gear don't send a backpacker to get comfort items. We send Penguini to get tents, and he comes back with six tents that looked like a condom to fit a body. He also buys one six-man tent to store equipment and use as our working area when we are not on the rubble pile.

There have been reports that several aid workers have been killed by criminal elements, so while our over watch is busy making liaison with headquarters, the Beast and I make connections to insure that we aren't unarmed while we are there. We have to be careful because we don't want to get our manager in the benjo pit, and the company in hot water, so we go by the old SF theory of it being easier to ask for forgiveness than to expect permission.

We have been trying to get landing permits to land in Haiti, using contacts we have in the State Department. We might have had better luck pissing up a rope for all the good they do. Bureaucrats are the same no matter what they did with you in the past, when they were real people. We try to get on the relief flights into the country, we try to get on anything smoking to be there in time. You'd think that help would be a valued asset, but it doesn't work that way. The government has a plan and we aren't part of it. The Penguini even tries his contacts with the NGOs; they are worse. Basically, if you aren't part of a registered NGO, with experience, they don't want you tarnishing their image. We would see more of this lassitude and do-gooder arrogance when we got on the ground.

In the end, Magic Mike becomes the jewel in the crown. We are in need of two important components, the first a private aircraft that is able and willing to land in the heart of chaos, the second some trinkets for our personal self-defense. The first arrives in the person of a gentleman who runs an airline that contracts for sensitive jobs around the Caribbean, who actually comes to us through the trinkets source. The Beast and I explain to Magic Mike that we aren't getting any help from State. He looks at us and says, "The only thing State is good for is excuses and the address of the nearest gay bar, what did you expect? I might know a guy," he muses. "Let me call him but let me warn you he is a bit odd."

"How odd?"

"Do you want help, or do you want to see his psych profile?"

No arguing there: if Mike thinks he could do the job, he's our man. The fact that he thinks the guy odd is actually a good reference. We don't need the lawful, we need someone that has a foot in the underworld and one on a banana peel with the feds. An hour later my cell rings and it's his guy, Tigran the mad Armenian, who is waiting in the parking lot. No sense getting the Hessian up for this, the less he knows, the less he can truthfully tell the authorities if it goes bad. Four of us head downstairs leaving Gator to keep the Hessian busy.

Mike had told us that if you needed anything in the Caribbean, Tigran was your man. He is a smallish, intense man with a lot of nervous energy and can provide us with all we need. He is a former FBI agent, who despite a sterling undercover career was determined to be too loose to keep on the payroll. Even the rest of the alphabet crowd are leery of him, suspicious that he would kill them all if he thought they weren't pulling their load. We love the guy at first meet, to him anything is possible. If we wanted a nuke or dirty pictures of some potentate, he would deliver. He tells us that the airplane folks will meet us in an hour.

We never leave the parking lot. He walks around to the back of his jeep and opens the door, whips back a blanket, and says, "I brought you a selection of what you asked for." There laying out like a display are four pistols, with a box of ammo for each, and a Heckler Koch rifle with all the frame furniture, night vision and 10 magazines fully loaded with spare ammo.

"I talked to my people and they say that the gangs are robbing everyone including the bodies. I figure you are going to need some heavy firepower. The mags are loaded with ball, tracer or armor piercing and the color tapes indicate which one. None of this is registered anywhere so don't get caught with it." I'm fine with that, the Beast is fine with that, the Penguini is trying to figure out how to keep all of it when this is over.

In the end we get the four pistols. I turn down the long gun, as tempting as it is. How are we going to conceal it? Not just from public view but from our proctor. Best to keep it small. We've been in bad places before when mob rule takes over and the rule of law is out the window. I don't want to be a victim and it's one thing to have to use a pistol to protect the team; it's a whole different case if you're carrying military hardware. Most of all I don't want to get our escort in trouble. The company wants us to use the technology to assist with rescue efforts, that is our primary mission.

The air asset shows up an hour after Tigran leaves. He is all business and brings along his chief pilot. Both are extremely knowledgeable. I go over our desire to land at the main airport; they in turn give me an update on the actual situation on the ground. It isn't good. They recommend a Lear, for its speed and capacity to carry us and our gear. I ask them if they can get an aircraft that can land on one of the highways on the outskirts of the city or in a field, if we can't get in on the main airport. The guy looks at me and says that for the money we are willing to pay for the charter, he will crash land in a cornfield and claim it was children playing with matches. They also say that getting clearances to land is a wasted effort. Apparently, State has their hand in clearing all flights in and unless you are on their approved list or someone from Hollywood, you're not getting it. They assure us they will put us on the ground despite of the restrictions.

* * *

We are to lift the following night and will be on the ground by 8p.m. We pack up and move the team to the private side of the airport in Fort Lauderdale. We pass through and load the waiting aircraft. It's a beautiful plane, obviously set up for executive travel. We are squeezed into the cabin with equipment stuffed in every available space. We are at the limit of the plane's weight restrictions but we lift off and head southeast towards Haiti. With the Lear we are faster than the commercial and military flights making up the airlift going into the country. I look out the window and back towards Florida. There is a chain of lights behind and in front of us like sparkling jewels in the dark night. Most of these are C117 military aircraft, loaded to the gills with relief supplies.

They are also getting ready to airlift the 82nd Airborne into the city to help maintain order and protect the relief effort. By now it is the third day after the earthquake. Just before we left, we got a call from our contacts in the State Department once again assuring us that they are trying to get us a flight. It's all bullshit, all that's left will be their apologies for not being able to help, after they get their second cup of coffee.

We are nearing the Haitian coast and the pilot turns and says, better buckle up we are about 15 minutes out. I can see that he has tucked us in between two C117s and slightly above them with our running lights out. As we drop into the slot, we start getting buffeted by the turbulence from the lead plane's engines.

We pass over the harbor on the approach. The city is dark except the few buildings that have their own generators, and the ships in the harbor. The buffeting becomes more severe as the C117 in front of us touches down. We can hear the following aircraft warning the one in front that there appears to be a small aircraft behind him; that's us. As we touch down our pilot makes a hard right onto the first taxiway, so violently that we think we are going to roll. He smooths it out and we come to a halt just before the terminal building. We pile out and start throwing our equipment into an impromptu stack beside the plane. The pilot secures the hatches, shakes our hands and tells us to call when we need a pickup. He taxis out and actually takes off from the taxiway and is gone in the stygian darkness.

We can't see bloody 10 meters it is so dark. We are looking for something to haul our equipment in. We find a bunch of shopping carts and load it into them. Gator wanders off in the dark. We decide to go into the main terminal and see if we can locate someone in charge. The first person that we run into is a French UN soldier that we awaken. He is surly and snooty about being woken up. His entire English vocabulary consists of "Go away," then "Please don't hurt me again." We find out that he is the only one in the building because it is heavily damaged. The whole place stinks of human feces, since they are using the hallways as a midden. We finally go outside and are looking for a place to hunker down for the night until daylight.

Gator shows up with a dilapidated van with the dimmest headlights I have ever seen. He had hotwired it in the dark and brought it over to where we are. We are loading our equipment in it when it gives up the ghost and refuses to start. We can make out a plane that is parked nearby and decide to sleep under its wing to prevent being run over in the dark.

Dawn breaks after we manage a few hours of sleep, ignoring the constant noise of military airlift on the far side of the airport and the overpowering stench of dead bodies on the air. Besides mosquitoes, the dawn brings us a view of the destruction wrought by the quake.

When the earthquake hit, the entire city lifted up eight feet and then slammed back down. Poor construction and the violence of that shock resulted in 300,000 people dying in the quake itself. Aftershocks continue to collapse those buildings that survived the initial violence. The Haitians had used very little rebar and in some cases, used aluminum chicken wire, bottles, and other scrap instead. Making matters worse, they had used salt water in making the cement, so it was brittle. The scene reminds me of Lebanon during the civil war.

We can see what appears to be an encampment of tents ranging from the standard military GP mediums to commercial types and hundreds of relief workers starting their day. I am wondering why they are here instead of in the city. We wander over to what appears to be the center of activity, and find the Relief Headquarters supervising the rescue efforts. In the distance, you can see a solid pillar of smoke rising; we find out later that's the pit they are burning the bodies in. There is no identity check, and no regard for race, creed, or citizenship. They can't bury them and they are already bloating, so they are spreading lime on the ones still laying out, and burning the rest as they collect them. I visit the site a day or so later as we made our way into the city and it looks like a scene from the liberation of the death camps at the end of World War II.

We find the headquarters tent, and manage to sign in on the board listing the rescue crews from all over the world. Our proctor goes inside with me. The information available is scant to non-existent, and I soon find out why. The effort is being managed by the UN, mostly French and a gaggle of Obama sycophants who were obviously not picked for their talent. The principal speaker is a thirtyish, scraggly female Georgetown graduate, who is yelling at an Air Force lieutenant colonel running the air operations. He had evidently done something that had upset her.

She screams at him, "You don't know what the president wants, I and only I know what the president wants. You will allow the planes to land that I direct to land, not your bullshit schedule." The colonel has the look of a man that wants to wipe the chickenshit off his shoe. Everyone else in the room is being silent as this raging witch berates him. I've had enough: I can see that the military is trying to do what it does so well, the colonel probably has 20 years of experience running air ops, and this dimwit has none. When there is a break in her screeching, I say from the back of the room, "Shut the fuck up!" She stops in mid-sentence and looks around the room. People part from around us until she can see who interrupted her tirade. I look at her and say that she should listen to the colonel, because she hasn't even a vague idea of what she is doing. She is so enraged that she may actually explode. We now have a clear area around us as people start slipping out through whatever exit they can find.

She points a finger at me and demands to know who I am. I tell her that I am Penguini, which really starts her up again. She threatens to have us thrown off the island. I look at her and say, "That's going to be difficult sweetie, because no one even knows we are on the island. Why don't you shut up and let professionals do their job." By now we are the only ones besides the Colonel and his aide still in the tent. He gives me a look of pity and gratitude and shuffles his paperwork. I quickly tell her that we are looking for someone rational so we can get to work saving any lives that are still in jeopardy, and that if she talks to me that way, I don't have any compunction in punching her out. She sputters and throws a "We'll see about that," over her shoulder as she storms out of the tent.

The Hessian looks at me and says, "Let me guess, that was your best diplomatic approach, right?"

"Yeah, and just so you know I would knock her out, she is obviously getting in the way, not helping the situation. If she is in charge this will be a real disaster. As far as I am concerned one more body into the burn pit won't even be noticed and judging by her actions there will be more than enough suspects that her disappearance won't even raise an eyebrow."

"Is that why you gave her the Penguini's name?"

We find out that the ruckus was over the fact that the Colonel didn't want to let some Hollywood ghouls and the drive-by press land ahead of relief supplies.

We go back and gather up the team and find a space between a Jamaican Engineer unit and a FEMA encampment with air conditioned tents housing FDNY. We spread out to find out what the true situation is and find out that it's a total goat rodeo. There is no organized command structure. No one is allowed to leave the airport until the 82nd Airborne gets here. The vast majority of the rescue groups are clustered in tent city with no water supply and limited sanitation facilities. A few have managed to get out and about, but the city is a dangerous place and not only because the buildings are constantly collapsing.

There are looters and gangs taking advantage of the situation. The day before we arrived six aid workers were attacked on the highway coming from the Dominican Republic with three of them killed. There are crowds of refugees fleeing the capital city. It is, in short, catastrophic. There is no medical help except the Cuban hospital in the city and those clinics operated by aid groups. The water system is broken and people are getting diseases because of it; add to that the numerous corpses still buried in the rubble and lying on the street.

The Hessian and I attach ourselves to a German rescue outfit, oddly run and funded by a man that had made money as a contractor in Iraq. We also spend some time with a Mexican group. The Beast and Penguini team with a South African crew. They are all well experienced and just as dedicated. These are the people that we will work with. Now all we have to do is get off the airport and into the city.

The UN has a fleet of trucks that were pre-positioned for just such situations. Being the UN, they installed security devices that were keyed to the biometrics of each driver. When the earthquake hit all those drivers died. There is no way to bypass the system. We know: we tried, unsuccessfully. So they sit like a huge mass of useless equipment.

There is no drinking water or a wash point for the rescuers. It is worse outside the wire, and there are mobs of angry, desperate people constantly outside the gates. There is an Ecuadorian armored vehicle guarding the main gate with weapons pointed at the crowds and it is forbidden to go beyond that point.

Magic Mike comes to the rescue here. Up to that point unless he had coffee, he was pretty much a passive observer. He had been here during the Papa Doc years, and has a network of people outside the wire. We go out the gate together, facing the mob trying to make contact with anyone that is able to help us get to the city. We are standing there when a voice calls out, "Mr. Mike! Mr. Mike! It's Mr. Mike!" Soon three figures make their way to the front of the crowd. He speaks with them and comes back to tell us that we will have trucks in about a half an hour. We assemble the Germans, the South Africans, and ourselves and split the devices so we can go to sites looking for survivors in the rubble. Soon, two dump trucks and a garbage truck show up and we all load on board. We will go to the National Cathedral and search on the way.

The second group will go to the main marketplace. The Hessian and I, along with the German team, dismount and start walking the last six blocks to the target area. There are stripped corpses lying in the street, bloated and covered with flies. People are wandering past with the stunned look of disaster on their faces, numb to all but their own suffering. There are way too many to assist medically. We find a survivor buried under a stairwell using the devices. We find another in a void of a collapsed building. We are going slow, but the machines are doing their job.

We have a dog with us that goes in after we have located a signal: he will dig down and find the victim then the Germans will do their magic. The girl who is his handler works the dog as if they have a mental link. I am concentrating on the device but look up as a young man wearing nothing but football pads and a helmet comes roller skating by between the rubble. It's incongruous. We get to the Cuban hospital and it's a charnel house. You can hear the screams from blocks away. They are harried but still working, so we make plans to bring any injured here. There are people laying everywhere on the grounds, with grievous wounds.

We turn back after several hours and return to the airport. I am exhausted and so are the rest of the crew. We stink of the dead and are filthy from scrambling around in the ruins. I have seen too many dead people and need to rest. There are no wash facilities to decontaminate.

The Beast and I go to the FEMA camp with the nice white tents. We are ushered inside. I ask for the man in charge. He comes up and right away I recognize that he is just this side of a pompous ass. I ask him if we can use their showers. He gives me this long-winded speech how that won't be possible because then everyone will want to use them. They are squeaky clean while we look and smell like a dung heap. They have a kitchen, showers, TV, cots, clean sheets, all inside their air-conditioned tents. I want to kill him and all his little girlfriends. There are a couple of guys that are actually firefighters who plead our case, but the anal bureaucrat refuses to budge. I find out later that management is made up of retired city workers, given the best of the equipment, and sitting on their asses giving interviews then running back to escape the heat.

By the time we get back I am beyond rage. I had told the posturing little bastard that he was a disgrace. That didn't bother him at all. When I get back the Penguini has made contact with the sergeant-major of the Jamaicans. They too have set up a water station and give us permission to use their latrines and shower point. We are able to wash the filth off finally and it is already dark. I am trying to buy a can of fuel from the Jamaicans, to go back late at night and burn the FEMA assholes out of their tents, when the Beast catches me and pulls me away.

I had developed a case of the shingles the day we landed and am in terrible pain. I am treating it best I can, trying to keep the infection from getting out of hand. My usual surly attitude has gone downhill and wrecked at the foot of homicide. I end up sleeping in the equipment tent as getting into one of Penguini's human condom tents is out of the question. Sitting, lying down, standing, leaning, all create excruciating pain centered around my groin and lower back. Only dying seems to be an alternative. Gator and Penguini have found a huge supply of water in liter bottles, stacked on pallets in the UN holding area, guarded by a lone French soldier. There is only one water point for the 1,800 relief workers and the line is constant.

Later that night, someone steals a fork lift after knocking the soldier out and duct taping him to his chair along with his gun, then drops a pallet of water in front of each tent occupied by relief workers. The UN is incensed and the next morning they are pointing and pouting about this lawlessness. The Jamaicans just beam at us the next morning and give us the thumbs-up every time we pass. Someone must have mistaken us in the dark.

That night the 82nd Airborne arrives. They come in, get set up and start sending patrols into the city to survey and accompany rescue workers. We are already loaded when a squad is assigned to us as security. The young sergeant is immediately adopted by the Penguini and we are soon moving like a well-oiled team. We finally get to the Cathedral. We clear everyone from around the rubble pile of what had been the Residency across the street. The Cathedral is in rubble with two partially standing walls left among the collapsed wreckage of the roof. The edifice with its soaring gothic face is tilted and in ruins, with only the center of what had been a magnificent stained-glass window hanging precariously at an angle. Only the face of Jesus is still intact.

We begin our search. It takes about an hour, but we locate at least five people still alive in the rubble and three more in an adjacent building. We are getting set to do a deeper search. The girl sends in her dog and it locates two that are buried shallow and the crews begin to dig them out. While we are standing there, we see a cameraman filming a news person. We walk over and it is a well-known CNN anchorman in his slightly smudged Gucci T-shirt and cargo pants. As he is talking three youths emerge from the rubble with a fourth who has a crushed foot and obvious fractures. He is moaning and screaming alternately. They lay him on the ground then run off to find something to act as a stretcher. As soon as they are out of sight in the rubble the scion of the New York press motions to his cameraman and tells him to make sure he gets the shot. They film the boy writhing in pain, with the anchor droning in dramatic monotone as he moves closer.

He reaches down and lifts the boy by one arm and drapes it over his shoulder. The kid screams in pain and yells at him to let him down in Creole. Mike yells at him to leave the kid alone, but America's News Face is not to be discouraged and hefts the boy up all the while staring into the camera, his heroic visage poised in the lens. About that time the other three come back with a door, see the whole scene and run toward him, yelling that they will kill him and skin him when they get there. He drops the kid and flees down the street leaving his cameraman behind.

One of them throws a piece of rubble as he disappears down the street. Mike steps in and saves the cameraman. They load their friend on to the door and make their way to the Cuban hospital.

I wish they would have caught him, I would have gladly filmed the whole thing as they stomped his smug face into mush. He wasn't the only one, we saw other Hollywood luminaries strutting around playing hero, soliciting donations on the tube, being the self-absorbed slime that they are.

We are with the Germans on one side of the ruins and are digging down towards where the signals are coming from. It is hard, hot work with concrete dust in your mouth and nose suffocating you. They have reached the basement and call up to the surface that one of us needs to crawl down to them. It turns out that they dig a path that can fit an international-sized stretcher. We had picked burlies as the team for physical stamina and if we got into a tight situation. I am the only one that can fit down the abyss. Rats.

The Hessian is too big by far, you would need a backhoe to dig a path big enough to shove his sausage-fueled body or the Penguini's down the hole, The Beast has just been pulled out and is puking his guts out beside the hole. He looks up at me and grins through the grime and says, "Looks like you're it Buckwheat." I take the device and start forward. I am claustrophobic as one can get, and the fit is tight. I am fighting the terror of being in the hole, every minute it shifts and dust cascades on me. I make my way down and pass two of the Germans and squeeze past them. I come across the decapitated body of a young girl, her head separated and sitting on top of the dirt. She looks like she is asleep. The cloying sweet sickening smell of death is everywhere.

The Mexicans, meanwhile, have located the body of the bishop. He is still warm to the touch. He had been drinking his urine to stay alive. It is hot and damp, sucking the moisture from your body. I pass a void on my right and there is a typewriter, undamaged and sitting on top of a half-buried desk. I swear I can hear someone singing spirituals up ahead of me. The rubble is compacted tight and the floor of the second level is inches from my face at an angle.

The whole pile begins to vibrate and everyone expects it all to collapse on us. The dust intensifies. It continues and we can hear a rumbling sound above us. Madame Secretary had chosen that moment to have her helicopter hover over the ruins so she could get a better view of the rescue effort. The helicopter almost brings the whole heap down on us, until finally it takes her posturing ass to the embassy. I had been unconsciously swearing in every language I could remember. As the vibrations stop one of the Germans coughs and looks at me and says, "That was colorful." Easing the tension. I use the device to determine where the signal is in relationship with me. The crew begins to dig again, very careful and coordinated. I am in the way, so they send me to the surface. It takes twice the time to crawl out, choking all the time. I can still hear a woman singing spirituals. It seems surreal. We are soon exhausted, and they replace the Germans with a Polish crew. These are big men and come with tools and big prybars and jacks to move the heavier beams and shore up the shaft.

As I came back into the blessed light I crawl out of the hole and am lying half crouched on a concrete pillar. I look over and the Hessian is being interviewed by a news crew. He is spotless and looks like the Rescue God. I want to throw up on him. I later see the news clip and he is the epitome of professional, explaining what we are doing etc. In the background is what appears to be a concrete gargoyle. That is me.

The sun is setting when the Mexicans bring the survivor to the surface. It is an elderly woman, who has survived long after she should have perished. They bring her out on the stretcher and she looks at me, she is so serene. They get the stretcher on the ground and the sun's rays illuminate the face of Jesus that had survived, and shine down on her. All the Mexicans are crossing themselves. It is that moving. She will be the last survivor pulled from the rubble. The others perished before we could get to them. She had been in the restroom and had a source of water in one of the toilets which she rationed for herself.

We come back with a sense of accomplishment that cannot be described. I manage to shower that night and my shingles has subsided somewhat. Magic Mike has located several cases of beer, some Coke, and some bottles of Barbancourt, the local Haitian rum, as well as some ice. We share a bottle of rum and a case of beer with the Jamaicans in thanks for using their showers and the friendship they extended to us. We expect to go out tomorrow but there is little hope that we will find anyone alive.

The next day we are out again. We go to three sites and we find no living survivors. Miller and Penguini go to a collapsed school with the South Africans. They are excited to find life signs under the rubble but when the rescue team emerges after exploring the site they tell them, "Well that gadget of yours works, unfortunately it wasn't survivors it was a couple of looters sitting around their swag pile."

The company decides to pull us out. They are sending our ride which will pick us up early that evening. We break down our camp and take the tents, stoves, ropes, gear, cooking stoves and rations over to the Jamaicans and donate it. The South African rescue unit is just five tents down from us. We have gotten more beer and ice from Magic Mike's reasonably honest supply chain, so we give the Jamaicans the rum and Coke, then load the lager onto a small utility vehicle and go down to where the South Africans are asleep on their cots. It is hot, and they have the tent sides rolled up. No one even looks up so we decide to leave them the beer as a gesture of goodwill. As the Beast and I are unloading the cases, the bottles clink. Suddenly, the shadows lengthen around us. It's all of them, awake and drooling in anticipation. Never get between South Africans and a source of lager. They're worse than hippos at a water source. One rings out, "My God the Yanks have lager." We barely escape the stampede. They are probably worshiping the Beast as some sort of Primal Beer God to this day.

I tell Gator to walk out around the grounds and spread the ammo we have since we can't take it back with us. He comes back in record time. I ask him if he got rid of it and he just nods affirmative and grins like the devil. I ask him where and he says that he dumped it in the slit trench that the FEMA twinkies had been forced to use after their super-duper latrine system failed the day before. Since they will have to burn it on the morrow, that is going to be interesting. I hope one of the loose rounds gets the woman from Georgetown as karma.

We wing back to the U.S. and split up after getting kudos from the company. We had proved the devices work, and will work as intended. We never get to deploy, as they have another solution based around armored ninja turtles, in convoy protection. A few weeks later, I write to a supposed friend and vent my bile at the relief scam. How useless the UN is, how vain, and a host of other small observations. He in turn blasts it on the net and I became the most hated man in Christendom. Every Liberal whiner and a few of the NGO crowd lambast me as inexperienced, uninformed, and a host of other adjectives. For them I have nothing but contempt. Millions of dollars of relief funds disappeared, no housing was ever built, the NAACP never adopted one orphan, and Madame Secretary went on to dig herself deeper into allegations of corruption and conspiracy.

For the rescue crews that donate their times and lives, you have our utmost respect. You are brave men and women and deserve better than you get from society.

CHAPTER NINETEEN

On the Road to Mandalay

Miller's son gets him a motorcycle for Christmas, just what every 63-year-old man needs. It is a custom bike from Johnny Pag and Miller is anxious to get it out and put it to the test. After a couple of months of riding around the neighborhood in Lake Forest getting used to it, he decides to take it down to Laguna Beach and then up the Pacific Coast Highway to Long Beach to attend a meeting of the Special Forces Association. At that particular meeting, a recently retired colonel gives a presentation on the mistreatment of the Karen minority in Burma (or Myanmar) and their precarious military position. He mentions that former Special Forces NCOs are being asked to volunteer their time to go over to Burma and help out with training and advice. Through some combination of altruism and boredom, the Beast winds up throwing his name in the hat as a volunteer.

At this time, Miller and the Penguini have just become roommates in a nice four-bedroom suburban home down in south Orange County. When Jeff gets back to the house, he talks Penguini into volunteering too. Penguini is 67. What could go wrong? They actually get a friend of ours to make a contribution to the non-profit that is sponsoring the program that covers their airfares. They call me and try to get me to come along, but I am already working on the same project from a different angle. So, one way or another, Burma becomes part of all our lives.

The Beast and Penguini fly over to Bangkok and then on to Chiang Mai, Thailand to link up with the support group for the Karen. Upon arriving in Chiang Mai, they check into a pretty nice guest house, right on the river, and wait for the Colonel to arrange for their transportation into the war zone. After a couple of days, they are taken to a safe house in a smaller village west of town where they meet a young British guy who is working for the foundation. A night or two there and they are picked up by a pickup truck to drive west to the Salween River. The trip to the river is amazing: it starts out on nicely paved highways, descends to narrow roads, to dirt roads, to stream beds, and eventually to driving through shallow water across flooded fields until finally coming to the river. The scenery is breathtaking: heavy jungle with just an occasional tiny village. Elephants, water buffalo, and thousands of birds.

They reach a designated stop on the river and eventually a boat shows up. It is a passenger ferry with two rows of seats down either side, about two thirds full of travelers. They lay the Beast and Penguini down in the bilge between the seats and cover them with tarps, push off, and continue up-river. Apparently, the Burmese military maintain spotters along the river, as well as helicopter patrols. One couldn't have two white-haired and bearded gringos being seen and reported. After an hour or so the boat pulls up to a tiny patch of sandy beach and our two intrepid adventurers are hustled out, up a short, steep hill and into a village of bamboo huts well hidden under the jungle fauna. They are welcomed by the Colonel, the Brit (who had taken an earlier boat) and two generals from the Karen National Liberation Army.

The village turns out to be bigger than it appeared at first sight. The portion where they land is a group of about six or seven huts, bordered on the east side by the river and on the north by a tributary stream that had cut a deep gorge over the years so there is about a six- or seven-story drop along the north edge of the encampment. To the south and west, extremely steep hills rise straight up from the backs of the huts. Later, they discover that at the northwest corner a bridge crosses the tributary and another section of village about twice the size is located on the north bank. A fairly impressive location. Miller and Penguini are assigned a hut in the first section, as soon as the Karen get a cobra out of the roof thatch who had apparently taken up residence while the hut was unoccupied.

They settle their small amount of personal equipment into the hut that will be their home for a while, explore the area, meet two translators who will be working with them, and have their first meal. Almost every meal they eat is the same: noodles, maybe with some vegetables tossed in, and topped by a fried egg. It is nutritious, not bad tasting, but after a couple of weeks or so, really boring. The Karen take some pictures and effusively welcome the boys, they were so appreciative of just the idea that anyone would come from America to help them, it is quite touching.

The following morning, everybody gets together in the largest classroom north of the stream. The generals and some other Karen talk to the assembled students and then Miller and Penguini introduce themselves through their assigned translators. The group is broken into two sub-groups, the first stays with Penguini in the main classroom to spend the next two weeks learning man tracking and the second goes with Miller back across the bridge to a smaller classroom to spend two weeks learning about raids. After two weeks they switch. The Karen soldiers are woefully underequipped; they have some excellent natural capabilities in the jungle, but little formal military training, and for the most part (with a few exceptions) they are just too nice.

The Karen are a Christian minority in Burma, which is one of the reasons the Burmese are determined to wipe them out. They are a hill people, totally at home in their environment and very friendly, almost gentle by temperament. So Penguini decides to add a joint class on self-defense techniques every afternoon in order to try to cultivate a little aggressiveness and a little fire in the belly. The troops love it. It doesn't take long to fall into a routine. Up every morning, boil some water for instant coffee, and then off to the training venue. Break just before noon and head back over to the hooch where the first meal of the day is delivered, followed by an hour of free time which they usually use to head upstream for a makeshift shower. Then back to work until about 4p.m. when both groups get together for their self-defense classes, then dinner, and sitting around reading or just bullshitting until bedtime.

Up the tributary stream about 200 meters is a waterfall, where the Karen have stuck a length of PVC pipe into the flow and turned it into a natural shower. It works constantly, sending out a stream of very cold water with the power of a firehose. Very refreshing if kind of hard to use. They have to climb along the bank of the stream up a very steep hill to the site and then strip down and wade out to the pipe. Get wet, shiver, soap up, rinse, shiver some more, towel off, and then climb back downstream to the village. Still in the jungle heat and humidity it is worth the daily trek.

People come through the camp almost daily, usually Karen military or politicians but occasionally outsiders. The Karen are experimenting with game cameras to watch the approaches to their territory and a couple of Europeans come to help with that project for a few days. Every now and then, someone from the Free Burma Rangers further up-river drops by, and then one day about three weeks after they arrive the "press" show up in the form of a pudgy, supercilious, self-proclaimed journalist who writes books and articles about visiting dangerous places. He arrives one afternoon, stays one night and leaves the next day, probably back to the dangers of Pat Pong Road in Bangkok. He comes over to hang out with Miller and Penguini in the evening. They don't identify themselves but let him do most of the talking, which he is more than glad to do. He claims he has just come from Afghanistan, where he has developed an intelligence network that is the envy of the CIA and MI5 combined, until this retired agency jerk named Dewey came and took it over and ruined it. Now this blowhard had no way of knowing the Beast and Dewey are very tight friends. Jeff used to visit his home in Escondido often and sit around among the totem poles and machine guns and of course, the framed presidential pardon, drinking red wine and smoking cigars and plotting. Dewey had been the "invisible hand" offering us top-cover on some of our adventures. Of course, Miller is too cool to mention this, so he just says, "Dewey, really? There's a name I haven't heard for a few years, is he still alive?" This, of course, sends the journalist off on a long rant about how his intel net was winning the war until Dewey came along and screwed it up with his ham-handed incompetence. Miller just nods sagely and takes it all in, storing it all down in his encyclopedic memory. Later, after they got back to the States, we were in Las Vegas having drinks at the Mandarin Oriental Hotel with the Spider Woman, Dewey's long-time partner and alter-ego. Miller recounted the story and we all had a good laugh about it.

Mostly, though, the days go by uneventfully as the Karen troops take more and more enthusiastically to the training. Miller's challenge is to impart to his students the need for speed. Shortly before Miller and Penguini arrived, the KNLA had done a raid on a Burmese military communications outpost. They had spent three hours on the target before they successfully destroyed the microwave tower. Now anybody that has taken Infantry 101 knows this is ludicrous so Jeff was pounding into them *Speed, Surprise, Violence of Action*, the watchwords for a raid right out of the Ranger Handbook. They learn to shout it in both Karen and English at the top of their lungs.

After teaching them the basics in the classroom, Miller has his students down in the flats by the river building scale models of Burmese military facilities to use in pre-raid briefings and moving through the jungle breaking off elements to set up ambushes to cover the withdrawal route from the target. The students are really picking up and moving with professional skill. This was their strength anyway, moving through the jungle like wraiths, being able to disappear into the undergrowth within a couple of steps of the trail.

Penguini, meanwhile, is schooling his group in the fine arts of man tracking—reading and interpreting the signs left by people as they transit the landscape. He had originally been taught the art in Vietnam by an old Cambodian tracker and had trained himself to be incredibly proficient, even writing a textbook for the Special Operations community in the late 1990s. He builds sand pits and lays down sample tracks showing the students how to judge weight, what the individual was carrying and how, casting for lost tracks, estimating numbers, etc., etc. They are both really getting in the groove. Even the self-defense training is going well, with the students getting more and more into it and showing increased aggressiveness every day.

The Colonel returns for a visit with a special mission for Jeff. A misinformation campaign with support from some interesting places is underway and they need some support, in the form of evidence of training taking place that supports the rumors. One of the participants is the publisher of *Soldier of Fortune* magazine, who is willing to do an entire feature; also members of Congress are willing to read information into the Congressional Record. So Jeff sets out to create a KNLA Underwater Demolitions Team, or SEAL team if you will. The goal is to create at least the illusion, if not the reality, of being able to interdict shipping in Rangoon harbor. He starts with some PVC, magnets, and a linear-shaped charge. The idea is to swim under a cargo ship in the harbor and use the four magnets to attach the makeshift mine across the prop shaft. The shaped charge in the longer piece of PVC will direct the explosion straight up, severing the shaft and disabling the ship. An article appears in *Soldier of Fortune*, accompanied by a picture of the back of the Beast's head with a drawing of the mine as placed over the ship's prop-shaft. It apparently does the trick, scaring the Burmese government for at least a short period of time that the capability exists.

In time, the first groups complete their training, and another group is scheduled to arrive. A graduation ceremony is held; the students are issued black T-shirts with a silk-screened Special Forces crest with "Free Burma" above it. A few days earlier, Miller and Penguini, who were starving for meat, gave some money to one of the Karen and sent him up river to buy either a pig or a goat. He arrived back just in time for graduation so the fat pig he brought with him serves as a graduation feast. The first meat they have had since arriving, with the exception of a couple of bites of Burmese jungle cat one of the Karen had killed about 10 days earlier. The Beast claimed it tasted just like the meat in a C-Ration meal, Beef Steak with Potatoes and Gravy. Made him wonder who had the US government contract to provide that particular meal.

Penguini has developed some kind of hernia; when he moves a certain way it looks like the movie *Alien* with the alien trying to burst out through his stomach. Since they have graduated the first group, he decides he wants to head back to civilization for some medical attention. The Brit arranges for a boat, and the two of them take off for Thailand one morning leaving Miller as the only foreigner in the camp with a brand new crop of students. He makes arrangements for a couple of Penguini's best students from the first class to stay and start teaching the man-tracking portion, while he starts on a new course in raids and continues the afternoon self-defense classes on his own—but not for long.

Just a few days after Penguini and the Brit have cleared out, Miller is giving a self-defense class and he is super intense; the students are actually a little scared. As soon as the class is over, he walks back over to his hooch, does a pirouette and a half and hits the ground like a set of car keys thrown out of a second-story window, out cold. A couple of Karen lift him onto an unused pallet in the hooch next door and there he lays, fading in and out of unconsciousness, sweating, shivering, raving in the grip of some unknown fever. It feels like Dengue, but that is impossible because he had already had Dengue in Malaysia in 1986. There is no medical care of any kind there in the camp, so they just kept feeding him water and Coke to keep him hydrated and let him lie there. After a time—one day, three days, a week? Miller wouldn't know, he was totally out of it—the Brit shows back up sans Penguini who has decided to return to the States. Needless to say, the Brit is a little shocked by what he finds. Training has been suspended, the Karen soldiers are just sitting around, and Miller is semi-conscious when he isn't completely unconscious. But it will get worse.

One night, the Beast decides he really needs to pee, so he staggers up from his pallet and tries to find the tree he normally visits for this purpose. Unfortunately, his fever delirium gets him turned around and instead of walking back into the woods he walks right off the cliff that leads down to the tributary and pitches head first out into space. He told me later that he was really proud of himself at that moment because his first thought was, "Well I've really become a detriment to the mission now." He tumbles forward, rolling through the thick underbrush, until his left shoulder hits a small tree growing out of the cliff. This spins him around and he winds up tangled in vines, hanging upside down about 30 feet below the top of the cliff.

The Karen, hearing the commotion of Miller's fall, think it could be an incursion up from the river by Burmese forces. They lock and load and rush to the cliffside to repel whatever invasion might be in progress. Of course, when they get there all they can hear is some dull moaning. The Brit shines a flashlight down and pinpoints the Beast, hanging there semi-conscious. Four Karen climb down and manhandle Miller's body back up to the top and lay him out on a tarp. He has a broken collar bone, two long and deep lacerations down his right thigh, and a laceration across his chest, all on top of his high (maybe 105 degrees) fever. In other words, he is a complete mess. In the absence of any real medical supplies one of the Karen produces a gallon of isopropyl alcohol and they wet down a couple of towels to clean all the blood, mud and plant material off Miller's supine body. This sends him immediately into shock.

They get the Beast somewhat cleaned up, lift him back onto the pallet and wrap him in a sleeping bag and blankets; there is nothing else they can do that night. In the morning he is still alive, so the Brit sends a detail down-river to a refugee camp where they have some medical personnel to bring somebody back to help. The following day a medic and two nurses arrive to see what they could do, which is not much.

The medic and the nurses examine Miller, attend to his cuts and bruises, but are helpless against the fever that is racking his body. They recommend he be evacuated to a hospital in Thailand. Of course, this isn't something that can just be accomplished instantly, so the Brit sends word to set up a boat and a reception team on the Thai side of the river, but it will take a few days to organize. Meanwhile, Miller is trying to fight his multitude of ailments and is pulling himself up for an hour or so each day and shuffling around the camp. One day he has to visit the latrine, which is just a deep hole inside a small hut, uphill a little bit from his hooch. It is a hill that would be almost unnoticeable during normal times, but now presents quite a challenge. Miller shuffles off, nobody paying much attention, but after about an hour they start to notice he is gone. The Brit goes searching and eventually his search takes him to the latrine. There he finds the Beast, pants around his ankles, butt up in the air, face in the mud, out cold. The Brit's first thought is that he is dead, but upon prodding the Beast moans and regains consciousness. Apparently, he had struggled to get up the hill and then to get his pants down, but the effort was too much. As soon as he squatted over the latrine he just toppled forward face first into the mud and there he had been until the Brit went looking for him. The Brit admits later that he was seriously traumatized by the sight.

They finally get the transfer arranged and a boat shows up to transport the Beast back downstream to Thailand. This time laying him down in the bilge is easy, he probably couldn't have held himself upright anyway. When they arrive at the landing site in Thailand it takes two Karen to drag him up to the road and throw him in a pickup truck. On the long drive to Thailand the Beast mostly hangs his head out of the truck and pukes; by the time they arrive in Chiang Mai he is basically just a desiccated shell. It is late in the evening when they arrive, so they go to the guest house where he and Penguini stayed when they first arrived. Miller passes out the second he hits the bed and the Brit scarfs him up the next morning and takes him to the hospital.

The hospital in Chiang Mai is clean and well organized. The waiting room seems to be full of older White guys with much younger Thai wives, with the practicality of women trying to keep their husbands healthy in order to keep those military retirement checks coming in. Miller and the Brit see a doctor and explain what has happened; a nurse looks over his external wounds but not much can be done. His shoulder was healing, just a little crooked, there is still a bump there today. His various lacerations are all scabbed over, and none seem infected. Then they send him to the lab for tests. They hang around in the waiting room some more until the test results came back to the doctor and sure enough, Dengue fever. Apparently, there are four different strains of Dengue and the Beast has now had three of them. The doctor writes out some prescriptions and they go over to the pharmacy to get them filled. They don't admit him, saying there isn't much more they can do and since he has survived this long he will most likely recover. He just needs to come back for weekly checkups. Best part, the entire bill for the doctor visit, the lab, the hospital, and the drugs is less than a grand!

They go back to the guest house and for the better part of the next two weeks Miller just sleeps and takes his pills. Finally, he starts to walk the two or three blocks downtown to eat dinner with the Brit. At first these walks are about half a block at a time with a long rest. Fortunately, in Chiang Mai there are comfortable easy chairs on the sidewalk pretty regularly. These are for foot massages. You just sit down, and a girl comes and gives you a foot and calf massage for about two bucks. The Beast doesn't want the massages, but the girls take pity on him and let him use their chairs to gather his strength for the next half block.

Slowly, day by day, he improves until he can walk the three blocks non-stop. Slowly, but at least without a rest. He finally decides he needed to visit a local barber shop as he is sporting over three months of unrestrained hair and beard growth. He walks into a local shop and sits down in the barber chair where he gets the first good look at himself since arriving in Burma. He is appalled! His face is thin and nearly as white as his hair, deep blue circles under his eyes and new lines that were not there before, his hair and beard bushy and unkempt. The Thai barber does the best he can and by the time he leaves the shop he looks almost human again.

Finally, after about a month they book a flight and the Beast flies home to sunny Southern California. He is still in recovery mode, but way better than he was when they hauled him across the river back to Thailand. I take one look at him when he got home and say, "Jesus Miller, I sent you for rubies and you came back with rabies!"

* * *

Christmas comes and goes, and sometime early the next year the Colonel who was running the program in Burma holds a conference in Washington, DC. The Colonel has aligned himself with a new organization run by a retired general and they have some big plans, so they are gathering up all the people they have worked with in the past for a meeting. Penguini is also invited but doesn't make it for some reason. One who does make it though is the Brit, Jeff had last seen him at the Chiang Mai airport when he boarded the flight home. They make arrangements to get together after the conference for dinner. At a nice Italian restaurant in the Virginia suburbs of Washington over veal and Chianti, the Brit looks at Miller and says, "Do you have any idea how much time I spent worrying about what the hell I was going to do with your body?" This sets Miller back a pace. "Really?"

"Yes really, I didn't think there was any chance you were going to make it out of there alive."

The next morning there is a surprise visitor to the conference, a former member of Miller and Penguini's A-Team from 30 years earlier who, along with his twin brother, left the military and moved to Southeast Asia where they had made themselves quite wealthy (breaking a law or two along the way but hey, you know). Miller meets the Twin that evening over at the Ritz-Carlton for a drink and the Twin drops a bombshell: he really wants to meet Dewey. Is this just a serendipitous event, or does he somehow know the Beast and Dewey are friends? In our business you never really know the answer to these kinds of questions. Miller tells the Twin to come to the west coast anytime and he will do his best to arrange a meeting.

Back in California the Beast's first call is to the Spider Woman, giving her all the details about the Twin and his brother and what they want. Dewey has unfortunately injured himself and is in the hospital but Spidey says she will meet and then we can visit Dewey in his hospital room if everything checks out. Miller talks to the Twin and arranges for him to come over in a couple of weeks. One adventure closes, but another whole can of worms opens.

CHAPTER TWENTY

Terry and Some Very Odd Pirates

There are few nicer places than the Montage Hotel in Laguna Beach, California. Built on cliffs above a small bay with breathtaking views of the Pacific Ocean, it boasts five-star restaurants and impeccable service. The Twin flies his Gulfstream into John Wayne airport and checks into the Montage just a few weeks after he and the Beast had left Washington, DC. The Spider Woman drives up from her hidden rancho in San Diego County for a couple of days of meetings. At this time, one of the big news stories is the outbreak of piracy along the Horn of Africa.

The Twin has some creative ideas on the subject, so after briefing Spidey on his operations in Southeast Asia he lays out his vision. Both Spidey and the Beast like it and they drive down to Escondido to visit Dewey in the hospital where he is recovering from a broken hip. They introduce the Twin and let them chat and get acquainted. The Twin comes up to the Ritz in Newport Beach the following evening and we all have dinner, Miller, Penguini (who had been the Twin's team sergeant years earlier), and me. It is old home week. The next morning, he heads back to Singapore with a promise from the Spider Woman to visit soon.

But first, she has other ideas. So, she and the Beast climb on a Lufthansa flight—first class of course, Spidey doesn't travel coach—to Geneva to meet with her friend Sjomann, a billionaire arms manufacturer and shipping magnate. Spidey thinks he will be very interested in the project and willing to provide financing. Spidey and the Beast land in Geneva and check into a small hotel about 10 kilometers east of town, right on the lake. The following morning, Sjomann shows up in a 300-grand Bentley twin turbo to take the two of them on a little tour of the area. They end up at a private cigar club drinking 50-year-old Cuban rum and smoking limited edition Cohibas. When they leave the club, feeling little or no pain, Sjomann throws the keys to the Bentley to the Beast and tells him to drive. He tried hard to decline, "Oh no, I've had way too much to drink to be trusted with this car," but Sjomann won't take no for an answer. The Beast manages to make it back to the hotel without incident, miraculously, and the next morning he asks Sjomann why he insisted on having him drive in his condition. Sjomann tells him that in

Switzerland the cost of a ticket is indexed to your net worth. "I'd rather pay your ticket than mine." Makes sense.

The two hang around Geneva for a few days, getting things organized, touring some of Sjomann's businesses, attending the Parade du Lac, which is kind of a cross between a college homecoming and a GAY Pride parade. While at the parade, Spidey stumbles upon a couple of Saudi princes, no telling what kind of princes they are because Saudi Arabia grows princes like the Arizona desert grows jackrabbits, but they are nice enough guys accompanied by their "personal assistant." Spidey sits on the balcony overlooking the parade route, exchanging pleasantries with the princes, when the Beast notices a group of young, hot women in short skirts and very high heels walking into the lobby below. The personal assistant immediately jumps up and runs out, returning in about 10 minutes with the women in tow. Turns out they are Moroccan hookers and the personal assistant is actually a personal pimp. Miller and Spidey remain just long enough to not look like they are fleeing the scene, politely say goodbye and leave. No need to get involved in whatever is about to go down in that room!

The Spider Woman is coordinating with the old man by telephone every night and finally everything is settled enough to head off to Singapore.

The Garden Wing of the Shangri-La Hotel in Singapore is incredibly opulent. Miller is starting to like traveling with the Spider Woman, a whole different experience from our usual fare of cheap motels and fast food made from insects. Everyone assembles in a conference area just off the lobby, Sjomann and some kid he has brought along who is supposedly a piracy expert, the Spider Woman and Miller, and the Twin. The basic plan: to purchase and outfit a fast, well-armed ship to search for and destroy pirates in the Indian Ocean off the coast of the Horn of Africa. The ship would be able to provide convoy security for cargo vessels or just range freely on the hunt. At that time there is enough nervousness and fear among the world's large shipping companies that they are willing to pay well for protection.

There are a lot of details to discuss, not the least of which is how to operate within the law. Most ports won't allow armed vessels that are not part of a national naval force to enter, so logistics is a huge factor. Then the operational costs and what kind of profits can be made, then of course the initial financing. So they talk, and talk, and talk. The meetings go on over dinner, over drinks after dinner, over breakfast the next morning, at morning meetings, over lunch, in afternoon meetings, always flirting with a final plan but never quite reaching one. Finally, on the third morning the Spider Woman is so frustrated she picks up a heavy glass ashtray in the middle of a screaming argument and wings it across the breakfast table. Luckily it doesn't hit anyone but creates quite a stir among the staff of the hotel when it shatters against the far wall. The Twin grabs Sjomann's friend and they stomp out; they came back about an hour later saying the Twin has found financing elsewhere and doesn't need Sjomann anymore. At that point, the decision is made to leave, and Miller goes up and packs. He and Spidey head for the airport. The Twin does wind up buying a ship, but never gets it operational in the Indian Ocean.

* * *

Miller wouldn't be gone from the Far East for long though: as sometimes happens another apple falls off the tree and hits him on the head just days later.

Back when we were doing the fantasy camp, in one of our brief periods of solvency and success, we actually had an accountant. She was a very nice and very efficient young lady of Scottish descent who for that brief period kept us as organized as we had ever been. She contacts the Beast one day with a simple question, "Do you still do the kind of things you used to do?" Of course, being a shameless opportunist, the Beast tells her, "Sure, if the right possibility comes along." They make an appointment to meet at a McDonald's in Lake Forest. The story she tells him is something new.

One of her clients does a lot of manufacturing in China. He was delivered a shipment of defective goods from one of his Chinese vendors, so he had refused to pay. The vendor, claiming the shipment was fine, sued him in a Chinese court for about 100 grand. Then the client made a big mistake: he went there to try to settle the issue in person. Unbeknownst to most of us naïve citizens of a free country, China has a law that forbids any foreigner being sued in a Chinese court from leaving the country until the suit is settled. So her client is stuck. To make matters worse it is all a scam, because he had actually settled the suit, paying 100 grand, and the Chinese vendor ran over to the court clerk and filed a second suit for 500 grand that same day. When he got to the airport, he was denied exit. Still stuck. He has been stuck there for over a year.

The accountant asks Miller if he could do anything; he says he doesn't know for sure, but he would be willing to do a recon and find out. The following day he is back at the same McDonald's where the victim's wife hands him the funds for the recon in cash, folded into the day's newspaper. I am in Africa, so he links up with Penguini to plan a recon.

Just based on a map recon it appears Taiwan is the natural extraction destination. The client is in Shanghai, but he can travel internally. They look at Hong Kong, but it has a set of negatives, not the least of which is considerable Chinese government influence. With Taiwan in mind they go to visit their visa stamp artist and order up both entrance and exit stamps from China and Taiwan. Penguini waits for the stamps to be made, while Miller heads over to start the recon. He lands in Taipei, heads south down toward Taichung and finds a cheap hotel. His first destination is Kinmen Island, a county of Taiwan just a couple of miles off mainland China. Kinmen is a fascinating place; it had been so heavily shelled by the communists during the revolution when it was claimed by the nationalists that the island is literally covered with shrapnel. Now the shrapnel is mined and collected and forged into high quality kitchen cutlery, prized by the world's greatest chefs.

The Beast takes the ferry boat over to the mainland and goes through customs, checking the procedures and getting a real set of stamps in his passport that he can copy and send over to Penguini, so he can use them to check the stamps being made. When he leaves, he takes a ferry boat straight back to the main island of Taiwan, and then grabs a flight over to the Matsu Islands to check them out. He loves what he finds.

The Matsu islands are a small archipelago in the East China Sea that constitutes Lienchiang County of Taiwan. They are a bit further from the mainland then Kinman, about 20 miles or so, but that could actually be an advantage. Even better, he hasn't been there long when he meets a young woman, Mai Ling, who speaks very good English and is quite helpful. The Beast checks into a small bed and breakfast on the island of Nangan, not too far from the airport. The next day he goes looking for a fishing port to check out the customs controls for boats putting into port after having been out fishing or even over to the mainland.

The island itself had at one time in the not too distant past been a complete militarized fortress. From the late forties into the eighties the tensions had been high; even in the 1960 presidential race between Nixon and Kennedy the question of whether or not the US would go to war to protect Kimoy (Kinmen) and Matsu had been a big issue. As a result, the island of Nangan is honeycombed with tunnels and dotted with forts. Most of the forts are decommissioned and many of the tunnels have been turned into whiskey distilleries because tensions had lessened, but the island still seems ideal for the Beast's purposes.

He calls Penguini and has him fly over to act as reception for a possible extraction and then heads to Shanghai to meet the victim/client. Going through customs at the Shanghai Airport with a set of Taiwanese visa stamps in his left sock and a set of Chinese visa stamps in his right sock has the Beast's spinal base puckered like the airplane seat on the way over had been made of alum. His rational mind tells him that the chances of being strip searched are all but non-existent, but what if… He makes it through customs and takes the Mag-Lev train into town to the client's hotel. When he arrives, he finds what can only be described as a cluster-fuck. The client's hotel room is full of people, and they are all talking about the situation and spit-balling possible courses of action. Before he headed over to China, Dewey had gotten Miller a briefing from some "experts" and they had told him every hotel room should be considered bugged by the government so not to discuss anything in the room. Well that ship had already sailed! The Beast soon realizes that working with an adult, especially an alpha-male used to being in charge, is going to be a lot different than working with kids. He finally gets the client aside, out of the room and without the crowd, and briefs him on his exfiltration destination in Matsu. He tells him they need to see about finding a bribable fishing captain who could make the run, to which the client replies, "No problem, I already have people working on that." Oh great.

Miller also inquires as to whether the client has had any contact with the US Embassy. The client shows him a printed list of the names and addresses of about 10 lawyers in Shanghai that the embassy recommended. That was all the assistance they were willing to provide. Many Americans think that if you get in a jam in a foreign country, the State Department will come to their rescue. Nope, doesn't happen that way. We have found that most employees of that erstwhile department are more concerned about you annoying the country they are assigned to than in giving any assistance to the American citizen in jeopardy. Back in our day there was an unofficial piece of advice passed out in Special Forces from the older guys to the newbies. If you get in trouble overseas, go to the British Embassy, you will get a lot more help there then you will from ours. We have always made it a point to stay as far away as possible from all embassies.

A day or two later, a Chinese former employee of the client shows up and the entire group decamps and heads down to Fuzhou. Apparently, a friend of a friend of this individual has found a fisherman who is willing to make the 20-mile trip for 20,000 US equivalent in yuan. This fisherman's home port is in a small harbor about 28 miles from Fuzhou. Before they depart, the client has his minions running around town converting cash. It turns out that the equivalent of 20 grand in yuan is a really big pile of cash. It fills a pillowcase purloined from the hotel room like Santa Claus's bag of toys in a Christmas pageant. Who gets to hold on to it? Why the Beast of course.

The entire group check into three or four rooms in a hotel in Fuzhou, rent a car and later that evening drive out to the fishing port. There are four people in the car, the client, his best friend who has flown over to keep him company while he is stuck in China, the Chinese intermediary/translator and of course the Beast with a huge pillowcase of cash in his lap. By the time they reach the fishing port it is pitch-black dark. They pull the car into a gravel parking lot next to some closed warehouses and three of them walk down to the dock to make the final deal with the fisherman. Leaving the Beast with this big bag of money alone in the car, in the dark, in a deserted parking lot somewhere on the coast of China.

The only weapon the Beast has is a small switchblade he had purchased at a street market in Shanghai. He holds it like a security blanket as random Chinese dockworkers and fishermen wander by the car in the stygian darkness during what seems like hours until the other three come back with the unwelcome news that there is no deal. They are heading back to Fuzhou.

They find out later that the intermediary was trying to get the fisherman to make the trip for eight grand so he could keep 12,000 for himself and his cronies.

The fisherman wasn't willing to come down that far so the negotiations blew up. Of course, since all these negotiations were conducted in Chinese the client had no clue at the time.

They head back to the hotel and try to figure out a new course of action. The Chinese intermediary is excitedly trying to get the client to give him another chance, the client's friend wants to try to make it into Hong Kong, everyone is talking at once. Miller is just listening, waiting for the commotion to settle before working out a Plan B. But the client decides to take his friend's advice and move down to Shenzhen to explore the possibility of making it into Hong Kong. This, of course, makes all the preparations in Matsu irrelevant, so Miller calls Penguini and tells him to close down and meet him in Taipei. Miller flies over to Taipei when the rest of the crew decamp for Shenzhen, intending to gather up Penguini and head for Hong Kong. But it never happens.

Shortly after Miller's departure, the Chinese intermediary shows back up with a new deal: the fisherman has agreed to make the trip and they have to head back out to the port. Miller gets this bit of information immediately upon arrival in Taipei, where the Penguini who has already bailed on Matsu is waiting for him in a hotel. The Beast gives the client Mai Ling's contact information and instructions on how to find the B&B, then rushes to find Penguini. There isn't time to get back to Matsu, so they will have to wait and meet the client when he arrives in Taiwan. When he links up with Penguini he discovers that he had already destroyed the visa stamps, figuring it was over and not wanting to be caught with the stamps in his possession. Another major hurdle to be navigated.

The client calls from the airport, and all Miller can do is to send him to the embassy to tell them he lost his passport and get some temporary documentation to get himself back to the States. Of course, he tells them the truth, and what does the State Department weenie do? She calls the Taiwanese government and reports him as being illegally in the country! The race was in. The client is rushed south to Kaosiung where an air ambulance is chartered for the emergency evacuation of a very sick tourist. The client is put on board feigning unconsciousness and the plane takes off for the Philippines and then on to California. The Taiwanese police show up almost immediately after he leaves. The only real lead they have is the manifest of the flight from Matsu to Taipei, and there was only one other foreigner on that flight other than the client. The police interrogate his friend, but he is cool enough to tell them that yes, he had seen another American in the departure area but he has never met the man before and has no idea who he was. The friend heads back up to Taipei and gets the first flight back to Los Angeles. Miller and Penguini wind up flying home coach on some little-known Chinese airline with seats designed for five-foot-six, 135-pound locals.

* * *

The project gives the Beast an idea, always a dangerous situation, because of some information he has picked up in China. Apparently there are over a thousand foreigners in the same situation as this client, mostly Americans but some Europeans too. The Beast decides that an underground railroad could be set up with enough planning, and at least a good-sized number of these unfortunates could be extracted, with great profits. I am back in the States from Africa, so the Beast recruits me to join him in this project.

Of course, this will take money to set up—where to go? When the fantasy camp was up and running, back in Chapter 16, we met an attendee with whom we had subsequently become friends. This individual and his partner had some very unusual business interests, and one of them was helping people who wanted to get out from under the oppression of governments and live life as a free traveler in the world. Yes, it can be done, but it is expensive to say the least. We think this business plan might be right up their alley, so we take them a proposition for up-front money for a piece of the action. They jump right on it.

Problem is, we need cash, and not only did they not want to produce the kind of cash we need in the States, there are laws about what amounts you can carry in and out of the country. So we wind up with a couple of tickets to Singapore, a few hundred bucks, and an address.

We land in hot, muggy Singapore and check into Raffles, a hotel that just reeks of British colonialism. There is even a plaque in the billiard room commemorating the shooting of a tiger that had snuck in one cool morning in 1927 and was hiding under a billiard table. Of course, Raffles is also where the Singapore Sling was invented so it is all but mandatory to head into the bar for a couple of these notorious drinks. We call the Twin, not seen since the famous Shangri-La ashtray incident, to check in with him and get the latest Southeast Asia gossip. We go up to a small restaurant that is part of the Twin's holdings there in town for dinner and SF bonding.

The next day at the appointed time we are downtown looking for the address we had been given, It turns out to be an otherwise unmarked door that opens directly into a stairwell. Up the stairs, then down a hallway to another door leading into a small, oddly shaped office where a girl is sitting behind a desk that just barely fits in the room. We introduce ourselves and she seems to know immediately who we are and gestures us into an adjoining office, only slightly larger, where a fairly young man, obviously local, is sitting with what appeared to be a first-generation currency-counting machine. He is feeding handfuls of US hundred-dollar bills into the machine and making notes after each count. When he finishes, with nary a word, he hands us a large pile of bills wrapped in a piece of paper indicating that it totals 30,000. We thank him and leave. Armed now with the funds required, we spend one more night in Singapore and catch a flight out the next day for Taiwan.

We check into a hotel down south of Taipei which has an incredible spa down in the basement. Super-hot whirlpool baths, ice-cold baths, saunas—heavenly after a week of travel. We only stay one night and the following morning we head over to Matsu. We check into the same bed and breakfast that both Miller and Penguini had stayed in during the earlier mission. It turns out Mai Ling is off island for a couple of days, so we take the time to explore the island thoroughly, looking for places a small boat could put ashore away from prying eyes and transportation routes from one part of the island to another. Nangan is not big, about four miles long by two miles wide, but it is full of hills and canyons, wave-carved inlets and rugged cliffs. In other words, a smuggler's paradise.

Mai Ling returns a couple of days later and we go out to the airport to greet her. When she sees the Beast, she looks like a deer in the headlights, pure fear on her face. We take her for a bite to eat and she tells us that right after the Beast left last time the island was inundated with cops from Taipei, questioning everyone, trying to figure out how the American had made it from the mainland through Nangan and over to the main island and then away without getting caught and who was helping him. She tells the Beast that the network he had formed, her included, would in no way be interested in ever helping him again. Not after the secret police equivalent of a rectal exam they had just been through. Looks like we have run face first into a dead end in Matsu!

There is one other possibility, a long shot but what the hell—we are all into this project by this point. A backdoor through western China and across the border into Kazakhstan. Of course, this Plan B has a lot of complications: travel routes, temporary housing, and maintaining a low profile in a much more sparsely populated area. But it might be a chance.

We fly to London and check into our favorite hotel, the Union Jack Club just across the street from Waterloo station. Only current or former non-commissioned officers in NATO military forces are allowed to stay there. The downstairs bar has a picture hanging of every winner of the Victoria Cross since its inception.

We head over to the Kazakhstan Embassy to get visas and ran into the bureaucratic shuffle. Apparently, the US is making it difficult at that particular time for Kazakhs to get visas to visit the US so the Kazakhs feel they have to make it even more difficult for Americans to get visas to visit Kazakhstan. We try to leverage our friendship with the head of all internal security in the country, but the Deputy Assistant Undersecretary for Paperwork and Procrastination we are dealing with is singularly unimpressed.

He tells us two or three days, so we fool around London for the appropriate time, meet with an old friend, take in the British Museum. I let the Beast drag me around to visit places he had frequented 20 years ago, we discover some great pubs, and then go back for our visas. Well, two or three days has turned into two or three weeks. We think it might have been because we tried to namedrop earlier but whatever, it becomes obvious that we aren't visiting Kazakhstan anytime soon. London is an expensive city and our funds won't last.

So we catch a flight back to sunny California and have to tell our benefactors of our failure. They turn out to be totally cool; the Beast tells them we will scrape up the funds to repay them for the failed investment but they say to forget it, it's worth it to them to just have us around if they ever need us.

CHAPTER TWENTY-ONE

The Heart of Darkness

Ah, l'Afrique! One cannot say thy name without the sights and smells wafting back through your memory. Having lived and worked there, I can say that Rudyard Kipling's "East is East and West is West" never had a more precise delineation. Africa consists of lands with exotic and strange customs, a great seething mass of humanity who are climbing out of the shackles of colonialism, without shedding the cloak of its cultural impediments. We have seen it at its very worst, and at its occasional lulls between vibrant growth and utter chaos. The few that understand its culture and how to move within it are the White Africans, who stayed after the colonial powers pulled out with varying degrees of grace. The British left an ingrained civil service, the French left lassitude, the Belgians sulked like the French, and the Spanish and Portuguese left their mark on the gene pool.

Our passion has always been to see what's over the next hill. This passion has led to some dalliance with derangement, which is synonymous to "African Adventures." You hope that you will have the deep, soul-searing experience of the good Captain in Conrad's novel, but you suspect that it will be more of a *Naked Prey* experience like Cornel Wilde, conducted while gasping in survival mode.

We have been lucky in having the Doc in London as a guide. Fifty years of experience dealing with the different cultures and fluent in the languages, he surpasses even the best published Area Studies. These have in turn led to associations that last a lifetime. He is one-stop shopping in regard to useful intelligence. He gives us a brief on both the geopolitical and the mundane chicanery in any given country. While the Chinese moved into the sugar daddy slot, when the West finally got religion and decreed that bribery would no longer be tolerated, they found a vast continent to direct the energies of their economy and strategic policy. The Europeans went Marine and business adapted. We, however, acted like every deal had a pecker in it, usually the host government. America became a junior league player in anything except oil and mining. But the biggies like oil and mining still built and operated billion dollar projects and made money, all the while cleverly fencing with the outright chicanery and corruption flaunted by the Europeans and the Chinese. While we were constrained by laws designed to torpedo your business before you leave port, the rest of the world looked on and shook their heads at our naivety.

That sounds a bit long-winded, but unless you have grown up elbow to elbow with the cultural gaps, you cannot begin to spot the potholes on the way. During our travels we have often split and gone our separate ways, then found opportunities to include our friends in ventures.

Our first venture to Africa is in helping an American mining company that is involved with diamond mining in Guinea, West Africa. American Metals is based out of Costa Misery in California. We had been introduced by Dick F., a venerable survivor of the Frozen Chosin during the Korean War. He had an association with the lead figure in the company. The man was brilliant but like all geniuses he had a few quirks. His morning ritual was to drink his own urine as a way to detox his body. Thankfully he swished with mouthwash, so the morning breath was at least tolerable. We have run into a number of people that utilize this homeopathic preventive measure. We never tried it as we have enough disgusting habits as it is. The Beast and I met with him and discussed the possibilities of acting as his security force around their concession near Boke. At that time, the Russians have a strong presence in Guinea and the president Monsieur Touré had thrown his lot in with them in return for their economic and military assistance. We begin to do an area study and the initial planning to facilitate securing the area for mining. The company has developed a gravity separation apparatus that allows them to separate the diamonds from the alluvial soil, much the same way gold is processed. The system uses very little water, which is a boon in that water was a scarce commodity in the area they are operating in.

Of more concern to us is the presence of illegal miners operating around the concession, often with the help of the military, who are taking a cut of the diamonds for providing security. There is another concession just to the north of ours that is paying the miners in palm oil and foodstuffs for diamonds and gold and doing quite well without having to lift a shovel to get to the bright baubles. This is an example of Canadian pragmatism. The artisanal miners dig what was referred to as "Booger Holes," which were 10- to 15-foot deep holes with questionable shoring, into the gravel beds in search of diamonds.

Our plan is built around the military helping us clear the property of the artisanal mob, then securing the area with a police force paid by the company. This of course falls apart before we got the papers spread out on the planning table. The military is making too much money to let us get in the way. We eventually make a deal to let a certain amount of diamonds fall into the waiting hands of General A and General B and everyone is happy. All except the small miners that find themselves without work or sustenance. We eventually work out a deal where we eliminate the Indian middlemen who pay the miners squat for their labors and establish a zone on the old watershed where they could work both the gold and the diamonds and we would buy them direct. It all sounds quite neat and packaged but only to someone that has never dealt with the avarice of the average African government. It takes us almost four months to work the details out and have a deal that can be enforced.

In Africa, big discoveries of resources such as rare earths, gold, oil, copper, diamonds etc., attract the migrant population like flies to a two-day-old roadkill. In many cases you will find workers that have walked hundreds of miles just to have a chance at working and feeding their families. Everyone works in Africa. There are no safety laws, there are no preordained rules on child labor or economic slavery. On a continent that has no safety net for the elderly, and no agricultural system that can feed the mouths, population explosion is only one drunken night of foreplay away. There are many children per family, since one in four will die before they are five and one in ten before they reach puberty.

It never ceases to amaze us that the NGOs, guided by compassion and their beliefs, never focus on the basic problems: lack of medical and economic infrastructure, and sustenance, would be nearly eliminated if they would encourage contraceptives. This, of course, is impossible because a barren wife is a curse to the average African male. The more children you sire, the better chance you have of support in your old age. It is also a major machismo issue. Even if you prosper, the next country over sees you have opportunity, and they vote with their feet until you have overburdened shanty towns next to every mine site within a year. Add to that tribalism, fractured social boundaries, and sloth within the bureaucracies and you have Africa in a nutshell.

Africa is still a wondrous place. Its vistas have breathtaking landscapes, its cultures have evolved and maintained their identities, mixing with other waves of immigration, and produced a robust and vibrant presence. It's more like a hornet's nest than a bee hive though. The Beast and I are sitting in our offices with the mining company, going over the area study books, and reaching out to friends that have taken the Africa Dip before. The first thing you learn is that the mere mention of West Africa is usually treated with derisive laughter, followed by the account of their horrendous experience there. If you are so bold to mention Nigeria, they hang up. Fortunately, Guinea is not on the hell hole list yet, but it's teetering.

We have decided to set up a war room, complete with maps, so we can better understand the logistical and security issues. We come up with the brilliant idea to call Jimmy T. He is an interesting character, a cross between sheer competence and willingness to bend, or outright break regulations to help us out. He has a position at Bragg that gives him access to both information, in the form of After Action Reports from teams that have trained or operated in the region, and maps. All kinds of maps. We call him up and relay our needs and he says he will get back to us. When Jimmy sets out to do something, it gets done. There may be wounded and charred remains along his path but you get that with real information.

We had asked him for 1:50,000 coverage of the country, planning on reproducing what we needed beyond that from this stockpile. He calls back in a few days, gets our address and a few days later a truck pulls up. It's stacked to the rafters with boxes of maps. They are all 1:25,000 maps. These will require 10 times the wall space to put up but they are maps of the finest quality. We call him back and lambast him over the scale, but he shrugs it off, commenting he can get the 1:50,000, but he will have to forge travel documents and a request with the office stationery. We figure the more complex Jim in the Schmooze is, the more chance of surfacing on someone's radar, so we decide to live with the larger scale.

Sometime after that, we discover the Russians, who have an aluminum mine nearby, have stumbled across a diamond pipe. The Russians are the eternal opportunists. They never tell the Guineans that they have hit the pipe and begin to extract the diamonds. In order to ship them out, they call up Conakry, the capital, and say they have had an accident and poor old Boris and Vladimir were crushed beyond help and they are shipping their bodies back to Mother Russia. They fill the bottom of the casket with rough stones then line the top with sheets and place other sheets around carrion, to appear as if it was a body, then ship it back. In a country with little or no refrigeration, the caskets are barely ever opened, and the smell discourages the curious. They have been doing this for over a year. The Canadians also know about it but didn't want to disclose this, as they are making a fortune buying stones and gold with palm oil, so the scam persists.

We get an exploration team in country led by what they call a bush geologist, which is polite for prospector. He is a rotund gnome-like individual who smells outback. He is a very knowledgeable man, who educates us in how it works on the ground when your plan goes asunder. We manage to get in country to do a look-see and our first real gasp at Africa. After two weeks, we are all yearning for a hamburger and a hot shower that doesn't include removing the pit viper in the shower. We return and start the planning for moving the company to the project area and beginning operations. This means educating the gringos on what to expect when they get on the ground. After our horror stories on the food and its preparation the company hires a Greek company to provide cooks. The guy they hire is a real jewel. He is married to a Guinean and lives in the diaspora sections of the capitol.

We have noticed some tightening of the security in country with the army on alert and our one visit to the embassy had left us with the impression that Foggy Bottom would be useless if something went bad. The president is gravely ill and they finally decide to ship him to the Mayo clinic for heart work. He dies on the operating table, and his brother seizes power. He had been under a modified house arrest before this happened and with the older brother out of the way, he moves in and sets up shop. The country is in chaos with drunken teenagers roaming the streets rioting and looting, and killing anyone that isn't in their tribe. The expatriate crowd hauls ass as fast as they can. The Indian and mixed-race shopkeepers are the victims of mob violence. The Lebanese move next door where they own houses and businesses and wait out the strife, collecting the insurance on their burning businesses they are watching on TV.

We are semi-lucky. Lucky, in that we only have to retrieve two people and none of the equipment is there yet. Unlucky, because an old team mate, Seamus, surfaces in the most odd way. After we return home, the three of us are at the Special Forces Association convention in Bragg. Everyone comes by and gives us thumbs up and a wink. Since alcohol is present, we ignore what we don't understand, but realize that it isn't because we are such sterling fellows. Finally, one of our playmates congratulates us on overthrowing the country of Guinea.

Now most of the self-possessed egomaniacs we know would preen themselves and envision their being enshrined as a legend. But we know that a rumor of that sort attracts annoying little folks in polyester carrying badges and subpoenas. We quickly track down the source. It's Seamus. During his tenure with the Tenth he was known as rumor control and apparently he hasn't lost his penchant for the bizarre. According to the rumor, both the Beast and I were present in the operating room at the Mayo, when the president expired. Woven around that were insinuations of the government being involved, British mining magnates, throw in a couple of Russian arms dealers, and you have the whole picture. We hear at least six versions of the lie. Hilarious that anyone would believe it, but the feds could make a ham sandwich into a banquet for 30 if they thought there was a promotion in it. We make a few select phone calls to dampen that enthusiasm.

* * *

We once again separate and go in different directions. The Penguini hooks up with one of his karate pals who is doing medical charity work in Africa, and is soon accompanying Doctors Without Borders to Sudan as their security chief. They are trying to vaccinate the children in the refugee camps. This means flying into Sudan and going to the camps. The Sudanese government flies over in the daytime and rolls 55-gallon oil drums converted into improvised bombs out of an Antonov 21 on the camps for grins and chuckles, while the Islamic militias raid for slaves and loot. There are numerous documentaries on the plight of the Christians and animists perishing by the hundreds of thousands, yet no outcry from the Black Caucus, or the two esteemed "Reverends," despite their prominent activism. I guess the victims weren't Black enough.

Penguini's last trip is cut short when they are leaving one camp, and just after takeoff the plane is hit in the co-pilot, making the trip terminate right then and there. He is soon back with tales of the camps, the countries and the NGO crowds that are there. Some are excellent; some are full of Euro-trash leftists, or our own home-grown version. All in all, it's like the frontier life of the 1700s. He actually saw a slave caravan from a distance, with the raiders pushing their slaves to the markets in the Middle East. Long lines of women and young children shackled together to begin their life of misery. There were separate lines of men trussed up the same, who would end up in the salt mines or lead mines to the north and west. No safety regulations there. They will die of dehydration, starvation, and inhaling toxic fumes before they are 10 years older.

There are happier times as well. I have an old friend, Ian, that was in Rhodesia, who has taken up residence in Nairobi and is in business with the emir in Dubai. He had both his legs screwed up when his Land Rover hit a mine during the troubles. He had to reinvent himself when the new government seized all his family's property when they gained the palace.

He and his wife have a shipping company that includes a fleet of Mercedes trucks and Indian drivers. He also owns a boom, which is a larger vessel than a dhow but similar. It has a shallow draft and can almost turn on its axis. It has both sail and two turbo-charged cat diesels and can make nearly 20 knots when necessary. He is transiting the East coast of Africa, calling on "nontraditional ports" usually at night in some estuary, loading and unloading at hasty wharfs or off rafts. He hauls rototillers, trucks, foodstuffs, diesel fuel in 55-gallon drums, prayer mats made in Italy, generators, freight, and an odd assortment of weapons and ammunition. He returns with gold, palm oil, spices, coltan, diamonds, ivory, and some exotic animals occasionally, all based out of the Emirates.

He needs a supercargo to keep the onboard books and be there to make decisions about freight, payment, and security of the vessel. I agree to come by and look the operation over. I am immediately impressed with the ship and crew. The ship looks like a tramp but the engineer is a German, the engine room is gleaming with oil and clean as a whistle. You could eat off the floor in the engine room. His two assistants are Ghanaians like the captain, and are accomplished mechanics. There are also plenty of spares for those little accidents like hitting a hippo or being rammed by some snag during flood season.

The captain is a bearded Ghanaian that looks and talks like Fidel Castro, who had in his youth been a Portuguese commando in Angola before he moved to Ghana because he had a price on his head. The rest of the crew are a mixed lot but share a commonality of being excellent seamen and could double as a fighting force if needed. There are hidden spaces where weapons and ammunition are stored on board along with an Oerlikon chain gun to dissuade boarders. They all work for my friend and are fiercely loyal to him. They have a system and routes, I only need to be on board to do the transactions, collect the cash or barter value and bring it all home to daddy. It seems simple enough. I have visions of being the rogue trader, sailing the seas with a scurvy crew. Reality is that it is a business with strict rules and operating procedures. Sailing these waters requires expert seamanship, an intimate knowledge of who can be paid off and dodging patrol vessels of six different navies.

Soon after I accept the job, we are on our first voyage out and laden to the gills. Our ports of call vary from a true port, to up some river in the dead of night with the tide. We are out beyond the 12-mile limit in the shipping lanes when a British corvette hails us and informs us that they are going to board and do an inspection. The captain is having none of that and heads our happy ship into some shoals where they cannot follow; we make it over the top, and they pace outside for several hours waiting for us to come back out. The captain knows another outlet and we slip out and back down the coast sometime before dawn.

Each stop is a new adventure. When we are at sea it is refreshing and monotonous, we are never far from the shore in case we have to make a dash for safety. This also puts us in range of a variety of pirates. Most avoid us because they know the emir is involved; the few opportunists are either warned off or reduced to splinters and an oil slick with the chain gun. I personally never haul ivory, since it is becoming too hot of a commodity and because I love elephants and detest the traffic in their teeth. But by that time more lucrative cargoes are coming on board. Coltan and the rare earths are fetching twice the money for half the space and more tonnage. But we haul a lot of cargo that you wouldn't find on a traditional freight forwarder. Depending on who you are delivering to you are either armed to the teeth or having a pleasant exchange in the main cabin, followed by a cigar and adieu.

I thoroughly enjoy the time. Ian is recuperating and wants me to fill in while he has another series of operations on his legs and back. I am happy, the crew and I get along like the feral little beasties we are. It's a fairly carefree life and I am happy for a change. There are certain individuals in Kenya who don't like me, one of whom is in the various naval forces tasked with stopping smuggling. He suspects that I am involved. He suspects that my friend and I are up to our eyeballs in the trade. I spotted him giving us the skunk eye when I first got there, trying to shadow us after seeing me with my friend. We had words at a restaurant, and I basically told him to piss off, but he keeps showing up either at sea or ashore. He even goes so far to have a chat with the embassy about me, which filters right back to me through the grapevine. He told them that I was the scourge of East Africa. The way he described it you would have thought that I was Jean Lafitte and his fleet. Either way he is getting to be a pain in the ass.

The Beast comes sauntering through town and we go out for a pub crawl. We are in our second or third pub with a couple of Rhodies and a Royal Marine in tow when Jeff leans over and gets my attention. "That naval officer seems really interested in us." I look over and there is the good commodore or whatever he is, in his tropical shorts and uniform. He is glaring in our direction and talking to two other White guys in civilian clothes. His face is all flushed and he is talking excitedly and gesturing towards me. Our eyes lock and I give him the finger. I'm in no mood for this. I turn to the Royal Marine and nudge him. "You might want to make yourself scarce if that mob comes over here." He looks up and blearily retorts, "I'm not worried, he isn't even wearing big boy pants," and goes back to his suds.

We watch as the group moves off and we get back to hydrating ourselves. Sometime later one of our contacts with the embassy shows up to inform us that the officer has been making waves about my presence and connection to Ian and his merry crew. He is in full tilt to have me ejected from the country. The Beast has made contact and become friends with one of the richest men in the country and says that he will have a chat with him to throw water on this annoying little pest. Later in the day, we go by a print shop and I tell the Beast that we need to stop. I go in and order 500 business cards for each of us. They are embossed with the company logo for Ian's shipping company and for title I have Scourge, and he has Plague.

We have a nice three days seeing the sights from various bar stools and squiring around in a Land Rover, terrorizing slow pedestrians. On the second day, we drive the Royal Marine down to his ship and give him a royal sendoff. Royal Marines are a handy tool when dealing with tight situations that require brute strength and tenacity. They are great slabs of meat with the attitude of a honey badger. I have given the new business card to everyone I met so they are in distribution. Impulse is a bad habit. I figure a sea trip might be a good idea so we sail the next day.

When I return, I am met by Ian and he ushers me into his study. It's the typical man cave we seem to construct, with the commensurate "wall of shame" with all his plaques and awards from his military service and a few trinkets that he has picked up since.

I can smell bad news on the air, since his wife scurries out of the den and he eases himself into the high-backed wicker chair behind his desk. He looks like Sidney Greenfield only skinny. He starts by telling me that he has loved having me to pick up the slack but since he is healed he will be taking the reins from here on in.

I am poleaxed. I ask him if this is an economic decision and tell him that it has been a great time but I would like a little time to find something else. As usual my savings program was supposed to start tomorrow, so I am short on funds. He explains that it is mostly that, and to emphasize it he hands me one of my business cards with "Scourge" for the title. He says that someone from the Kenyan government had handed it to him the day before and suggested that I might be a little too obvious to have around. In the end, he gives me three months' wages and we part laughing over the little snit who obviously was behind this. In the end I couldn't resist, and got bit because of my lack of decorum.

I have a lead on an oil company that has some troubles in Sudan so I am gearing up for a shot at advising them on their security issues with a friend of Ian's from the SAS. I decide to move to Addis Ababa. It's away from the prying eyes and snitch ears of officialdom and I love the city. It's one of the cleanest and most well-ordered in Africa, once you get away from the bustle of Nairobi.

CHAPTER TWENTY-TWO

The Lost Empire of Prester John

Two months into the contract, I get word from Nairobi that Ian's captain had been pursued into the mouth of an estuary by the snit. He had led the patrol vessel through the twisted channel and managed to get it run aground. The Brits had shelled the raft he was towing, and it sank in 20 feet of water. He had cut the line and made off in the dark. He called a ship in distress for the Royal Navy, eliminating the chance of the ship covering up their grounding.

In every navy the cardinal sin is to run your ship aground. Rumor is that the good commodore was relieved of command and sent to the Arctic to shepherd fishing vessels. Ian came back three weeks later and raised the raft and its cargo none the worse for wear.

My next contract has me living in a compound, and traveling up-country several hours by Land Rover to spend three weeks at various sites around a full-scale drilling program to establish the ore body in one concession, and oil deposits in another. It's a combination of drill crews and their support systems, and geophysical crews and their logistic chains.

I have a compound located west and slightly south of Gondar. The medieval city is home to some of the most magnificent castles built in the 17th century and near the lake that feeds the Blue Nile. It's the typical walled compound that you see in the back country. There is a small thriving town with a bus that comes twice a week from Gondar. We have set up an operations center where we can rest, re-equip and gorge ourselves before heading back up-country where the crews are. When I am not in residence the staff, consisting of the cook and her family, the maid, a houseboy, mechanic, and security guy keep the place tidy and secure. The local village is ruled by the elders and it is a traditional Ethiopian agrarian society. I am the one and only tourist attraction there.

When I first arrived here we had been down by the river and heard what we thought was a crying human baby. We had searched the rushes and eventually found a baby hippo crying piteously for its mother. The poor little sod must have wandered for some ways because we never saw the mother. I had wrapped it in a jacket and tried to rehydrate it. When we got back to the compound, we had begun the process of saving it. We found that it would stomach donkey milk with protein powder mixed in. We named her Pricilla.

She soon began growing at a phenomenal rate and within a few months was the size of a small cow, then the size of a Volkswagen. She was very affectionate and could recognize the sound of my Land Rover and would burst out of wherever she was grazing and chase me back to the compound. The Ethiopians were bloody terrified of her, and she was bloody terror if she caught one in the open. Pricilla only took a shine to the cook, who had helped me bottle-feed her. She would follow her around, grunting and making little cooing sounds when she was near. Everyone else was fair game. She was starting to be a bit of a concern. She was quite plump and sassy when I left on my last trip and as I return and am driving up to the compound, I keep looking for her to burst from the undergrowth with her usual abandonment. There is nothing.

I have been gone a month and as I turn into the gate, the mechanics, who usually greet me happily, keep their glances down and look extremely unnerved; the same with the houseboy, and the cook just bursts into tears when I spy her and quickly runs into the kitchen area.

I have been called back because my security guy killed the milkman. Let me explain. My security guy is a tall rawboned warrior from one of the nomadic peoples, who had shown up on my doorstep dressed in rags, proud as his people are known to be, and simply told me that he would guard my house and my properties. The only thing of value he had was a long double-bladed, two-handed medieval sword that shone with use and care. It was sharp enough to shave with. After a short negotiation we had come to an agreement. He would guard my house and I would sustain him and pay him, the equivalent of 20 dollars a month. My staff was aghast that I would burden them with this uncouth savage, and even more so when I had them clean out one of the tool sheds to make him an accommodation. This was shortly before I had left for up-country.

I had received a message both from my bosses, and from the Beast, who had arrived at my compound a day before. He told me that the authorities had my guard and that the place was in turmoil. I have bounced in the Rover for 15 hours getting back, but here I am and the first thing I do is find the Beast and to check and see how much of my liquor supply he has consumed. I find him in the main room of the house, sitting on the veranda watching the sun beginning to set in the hills to the west. He has a glass with a finger of scotch, that I am is sure the remnants of three fingers. He turns and looks at me as I shake some of the road dust off. The houseboy whizzes in with a basin and some water with a rag to wipe my face and hands. He too won't make eye contact with me.

"What in the hell is going on?" I ask him and pour myself a cold lager after fishing it from the supply in the cooler behind the "bar." Ah, that cuts the dust. I burp and wait for an answer. He stands up, which is a bad sign since when he is standing there is a long, detailed account in the air.

"Well," he begins. "Apparently your savage killed the milkman two days ago and is in the lockup in the village. They have tried him in the tribal council, and they are awaiting your return and the arrival of the army sometime tomorrow or the next day." I have drained the first lager and fetch another.

"You want to elaborate?" I ask him, and he gives me that look like I am a retard because the facts are clearly there. I look at him and ask him, "Where is Pricilla?"

He looks at me puzzled and asks, "Who the fuck is Pricilla, have you taken up with one of the locals?"

"Pricilla is my pet hippo you moron." He looks at his mitts and slowly enunciates that Pricilla had been part of the settlement.

"What do you mean part of the settlement?" He goes over and bleeds the scotch supply again and tells me that I should sit down because it would be better. I don't like this, not at all. Corporate is breathing fire down my back about my unorthodox living arrangement, I have been forced under duress to come back here to settle some domestic problem, I am hot, tired, and starting to feel uneasy.

"What are you doing here anyway?" I query. He just shrugs and says that he was just passing through and decided to come out to see how I was doing. "You have a shit storm to solve here you know, I am kinda under house arrest until you got here. Sort of a hostage to your good behavior." He muses.

"Then you should abandon all hope, because unless someone gives me an accurate account of what is happening, I am going for the big caliber guns and a couple of cans of petrol to burn down the village and get my guy out." He just laughs at me and observes that the village is already armed and ready for that.

I finally get the whole story out of him. I had left specific instructions with my bodyguard as to who was staff, who was permitted in or around the compound, and a few etiquette pointers for dealing with the people he considered as dogs, which included the village and everyone I had not ticked off the list. I had forgotten the bloody milk delivery guy. He brought me camel and goats' milk about every three weeks. Usually, the cook told him when I was coming back. I don't know how she knew because the company never called her, nor did I. He had shown up as requested and had strolled up to the gates and started in. The guard had stopped him with a long sinewy arm and told him to go away, the master was not home.

The milkman had taken this as an affront and had lambasted the guard for being a savage, half-ignorant son of a pox-riddled baboon and had tried to push past. He probably never heard the sword as it snicked out of the scabbard, whooshed in a long arc and cleaved him from the shoulder to the groin, leaving the head and half a torso standing for a moment, then falling to the ground. The villagers took affront and had finally smoked the guard out of the shed, finally subduing him after a few had gotten sliced up pretty bad. He was being guarded in the cement blockhouse built during the colonial era.

The widow was heard, the maid was heard, the rest of the staff was heard, and they had decided that since he was my man they would turn him over to the army rather than pour gasoline through the one small window and throw a match in. I had been assessed a fine for the dead milkman, and as a kicker they had hunted poor Pricilla down, as she was a chancre sore on good order. She had terrorized the population long enough. I had wondered why everyone had such a sleek look and fat buttocks, a sure sign of feasting—now I knew. Only the cook had refused the meat.

We finish off the night, and I am awakened by the official delegation around seven the next morning. The Beast and I go down and hear the heinous crimes against good order, haggle a bit over the settlement but agree to pay. It costs me about seven grand, most for the widow, and a small stipend for the cook who is still grieving over Pricilla. For several months after that, swains from across the land would arrive on the dilapidated bus in mufti, sandals, and a worn suit coat, sometimes a hat, to court the wealthy widow with her dowry. My relationship with the elders is shaky, but we both try and put it all behind us. They still suspect that some night I will go berserk and burn the hamlet down, but I talk them out of turning over my guard to the army and promise not to do the above if they agree.

When he is released, I drive him and his cutlery out of town in the direction he had come from. Thank him for his service and fidelity and tell him that he is free to go. He takes the great sword out, blesses me and strides off with the termination cash and his pride, back to the century he came from. Corporate gets off my back and a few days later I get rid of the Beast before he finds the secret stash of the good cognac and single malts. He mentions in passing that maybe corporate should give him something for his troubles, being the hostage and all, but he must have guessed that I was plotting on how to get rid of the evidence that he had been here, because he soon drops it.

* * *

Fast forward a number of years, I am back in the USA and have gained a new partner and protégé, while the Beast has settled in to his natural state of outcast from family and society, chasing gentlemanly pursuits like owning part of a signature rye labeled "Toxic Masculinity." He has partnered with a brain that makes Bill Gates and that sleaze who ran Apple look like pikers. This lad had grown up in the hard scrabble of a trailer park, and eventually transitioned from Donkey Kong to writing some of the most sophisticated software on the planet. He took the dot com route and finally sold his business to pursue "adventure."

My protégé is 20 years younger and a product of the Marine Corps, the Army Airborne, and investment banking. He is an energy force in his own right, and

I am comfortable in our efforts. He is also seriously twisted, but he is a sponge in sopping up situations and can meet with the president of a country in the morning, and swill beers with the factory workers, pimps, hustlers, criminals, and traveling minstrels later the same afternoon. I have decided to turn over my networks to someone who still has the fire in their belly for the contest. Julio is me with manners.

Julio and I have gotten involved in power projects in Africa. I fought the impulse to just open my wrists and bleed to death, the entire time. We started out in Nigeria with a software application that could secure bank accounts from phishing and other hacker activities. We had met and nurtured the inventor of said software for over a year. His name was the same as the favorite husband of fifties sitcoms. He was the most deceitful, weaselly, oily little gnome this side of a fairy tale. He had conned an equally back-biting set of investors into developing his product and now it was show time. We had decided to try his software in the most porous and infiltrated system we could find to premier the product. So we have packed off to Lagos, where we meet with an impressive array of figures in the government, banking, and development.

Things go awry as soon as we arrive. The Nigerian that had enticed us into the project turned out to be a prime example of everything negative you ever hear about Nigeria. The Cuban inventor has run up a 700-dollar bill at the local pharmacy for Viagra, and is having lap dances at the pool in the afternoon. We manage to get the application installed in three banks and operate it to demonstrate its economy and effectiveness. The whole idea falls apart as quick as we prove up the technology. The Cuban and his program writer get robbed by their former partners and we end up with intellectual property that takes a fortune to defend. So we switch gears and cast our lot to build small to medium power plants for commercial clients to replace the diesel generator banks so common in Africa.

We get an offer to present our products and services to the government of Cameroon. They have what seems to be a fairly stable economy and society, ruled by a despot, whose wife is one of the wealthiest women on the continent. Nothing going on here folks, just move along. I pull out favors from an old and dear friend who is a major player in one of the large mining companies. He happens to be an electrical engineer and has built major projects in Africa as well. He agrees to go with me and add some weight to our credibility. The trip is a nightmare, routed through Istanbul, where after arriving my luggage never shows up. It would take the office of the president threatening Turkish Airlines to finally deliver it to my home in California a month after I get back.

The Nigerian turns snake and my mining friend calls him out at the hotel bar in front of everyone he is trying to impress. It is quite a sight, six foot four of Ibo, spoiled mama's boy, trying to be threatening to my friend, who is a slight 160 pounds and sitting on a stool. He, in the course of his arguments, had called my friend racist, a defense he learned while studying at an American university. Shortly after that, he found out what a verbal evisceration was like. End result: the Cameroonians like my friend and hate the Nigerian.

We meet a lot of people in those two years. We meet the New Africa. That generation of Africans who are well educated either at their own universities (and there are a number of first-class education systems in Africa) or at universities in Europe and the United States. But if you are poor you are getting just the basics. The underlying problem of corruption will continue to stymie any attempt at improving Africa. When we think of corruption, we nominally assign it to one or two individuals, but it can permeate every aspect of society. In Cameroon for example, you have to bribe the doctor to treat you, he in turn has to pay someone in the hospital or in the health ministry baksheesh, or a bribe. You bribe the nurses to actually bring your medication, or you don't get it and if you are poor, you get nada.

You cannot break these chains with laws written to constrict business for one side and leave the other to continue to act in the same way they have for decades. Our intent was to bring real engineering and more importantly funding to bear on infrastructure improvement. My friend kept cautioning me to not be so optimistic, as there was always a horror show somewhere in every deal. He said to limit the liability, if not eliminate it altogether, was the only formula for dealing with Africa. I should have listened more carefully, but I was infected with Julio's optimism and the demeanor of the younger generation that wants change.

We have meetings with all the proper authorities, and my friend impresses the panel with his depth of knowledge as well as his forthright manner. The Nigerian spends the entire time trying to minimize our effort and redirect it back at his paper balloon. The end results are predictable: his first due diligence uncovers the fact that he is 99 percent hot air. Or as they say in Texas, "All hat and no cattle."

I have to go out and buy a suit in order to attend a photo op at the Palace. The only shops open carry a limited selection and it is probably second hand. Some car salesman in Milan discarded the clothing, it made its way through the rag market and ended up six steps later in the window of the shop we found near the hotel. I never wear it after I bought it, preferring to go in a polo shirt rather than look like some cheap car salesman.

In the end, we have to dispose of the Nigerian, but it takes weeks. His own relatives I believe pull his rope in, or more exquisite is the thought that they fed him to the crocodiles because he attracted too much attention to their other activities.

My old friend and partner in Mexico, Francisco, has a company that manufactures mobile clinics. We do retrieve a contract proposal for providing hundreds of these, mounted on trucks, able to navigate even in the wet season. The program is to be funded by a trust, which was part of a concession award, as a caveat to provide "humanitarian" grants to the country. These are common and help the government provide services, and have the appearance of being the work of the government.

We spend several months trying to get it from proposal to actual implementation. The money is real, the concession is real, and if you had been awake, you would have seen that the grab for the money would wreck the entire endeavor.

We still have interest in the energy market, Africa needs both the products we make and the services we provide. We decide to shift our efforts to East Africa. Julio has contact with a deputy in the health ministry in this country. They want to buy a fleet of medical clinics. We design a trailer that can be towed behind a Toyota Hi Lux with a king cab. Everything is cruising along nicely and for a change in a timely manner. We exchange paper between the ministry and ourselves, and we are invited to come over for the signing of the contract. We are to sign with the finance ministry and the health ministry since the government will pay for the program.

I send Julio as an advance party since it is his contact, and am in Mexico City ready to depart with the owners of the company. Julio calls me as soon as he gets situated. He is smart as a whip, but apparently hanging with me has given him an internal alarm system.

The day before we leave, he calls me with disturbing news. He had been to the ministry, in the guy's office, when the "Deputy Minister" started having a set-to with the agent that had brought us into the deal. The agent is a registered procurement company with the government and the ministry. The berk behind the desk brings up that we haven't paid the 15,000-dollar fee for the license, but that can be arranged when we arrive. The agent and his pals drive Julio back to his hotel.

His hotel is a modern European-style hotel. It's also isolated and has one way in and one way out. He tells me that they all want to go out to dinner and that even though it's late, they'll come by and pick him up. By this time my alarm bells are going off. I tell him to tell them he has diarrhea and stay in his room.

I am trying to figure out how to get him assistance, because this smells bad, really bad. The only recourse is to reach out to all your friends, for someone that can be effective, and more importantly, timely. I have friends in a neighboring region but they can't mount up and get there in less than two days. If you're in a jam, you need the right tool for the job and you need to know that it is going to arrive as advertised, there is only Spider Woman. You are going to get scuffed up, probably be forced to agree to some questionable future favors, but the woman has expertise that borders on the insidious.

She is back to me before I can have second thoughts. She has a local contact that has a similar background to me and is closely associated with the government. He will call Julio at his room. Julio calls me and says the man has made contact and will be there within the hour. He also told him not to leave his room or take any visitors. Of course, by this time Julio is in full Marine mode, mapping out an escape route, thinking about breaking into a room that isn't occupied. He is listening at the doorways on his floor looking for the right one, when he looks out the hall window which faces the front of the hotel.

He calls me and relays that he thinks the cavalry has arrived. I ask him how he is sure, and he says that a convoy of three trucks and an SUV, loaded with armed troops in camouflage uniforms, has pulled up outside and the troops are herding everyone inside.

I hear someone with that twangy accent that originates in Boer Land. Julio hands the phone over, the gentleman asks if I am Nick, then says we have your mate, not to worry.

I have to wait a couple of hours but Julio comes back up and relays the events. His benefactor is indeed connected and has brought special troops that are connected to the anti-corruption effort in the country. The deputy had fled the country by midnight, the agent and another had been arrested, along with a few of the hotel staff. The police and the ministry have had this particular hotel under investigation for some time. There had been a couple of Western businessmen that had disappeared and ended up as floaters in the lake. The only common thread had been the hotel. There were two separate kidnapping and extortion operations going on. He had been moved to kidnapping when they realized that we were coming over. That would make four chickens for the plucking. Only the quick action of our new friend had prevented it from happening.

Julio gets to spend a few days with our new friend, who not only is very competent, but is mad as Mad Mike and drives like there is no other traffic on the road; a truly interesting man, with the same pedigree as The Beast and me. It is all part of the brotherhood of Special Ops that is bonded by shared experience, war, and sacrifice. Just when you swear off Africa, new beginnings emerge.

Epilogue

So as through a glass and darkly
The age long strife I see
Where I fought in many guises,
Had many names—but always me.

—GEORGE S. PATTON

Don't call me Ishmael, though the tale has all the tragedy and drama of Ahab and his crew. The end of the trail seems so distant, when your life has been designed around living for the moment. One day you realize that you are on your seventh decade and you have been witness to historical moments that the newest generation has no grasp of. Try explaining to a *barista* at your local Starbucks that Pearl Harbor was not the beginning of the Vietnam War, or explain pay phones and party lines to a millennial. You will get the blank stare of the uninitiated for your troubles. We, like our forefathers, consider ourselves as a hardy generation, having weathered everything from a winter trip to the outhouse, to the advent of on-demand porno, with a few marriages and wars in between.

We are comfortable at our current age; every day is a good day as long as you're on the green side of the dirt. The best parts are the memories. We have lived experiences that others only dream of. We have met people who made history and have lived with the consequences of other people's deceit and mistakes. Yet we are still here, wiser if not smarter, and that cruel mistress called adventure, like the siren's call is always tempting, always abusing you when you give in.

We don't think of ourselves as old and we certainly are not frail, thanks to a life of rigorous exercise of the adrenal glands and hardtack for rations. We are no longer about to crawl over someone's wall in the dark of night with cutlery in one hand and hope in the heart, unless of course the right amount of money is involved, and the right companions are along for the ride. The perfect end for creatures like us is to be like Wilford Brimley's character in *High Road to China*, fighting the right fight against some warlord, improvising mayhem from the materials at hand and surrounded by enemies. Much more attractive than laying in your hospital bed, in some discount extended care home, in your own piss-stained sheets, hoping that it isn't enema Tuesday again.

What will the future bring? is the mantra of our brotherhood. Hopefully the biggest adventure of all is in front of us. The Spider Woman has relocated to Africa and is constantly encouraging us to join her for new and bigger adventures, we are tempted. From another source come whispers of big doings in South America with riches to be made. Also tempting. But maybe, just maybe, the phone will ring as soon as we are done typing this with an offer we just can't refuse. We will pack a bag and venture into the unknown to that one last great romantic adventure.

We have tried retirement, it sucks. We are still outcasts in society as we still don't fit in polite society without filters and chaperones. I like to think that I have done well with what I consider an average mind. The capacity for curiosity is the elixir of life, extending from your first brush with mortality, to looking over the next generation for possible protégé material, a select few of the SF kids, who have shown the same *joie de vivre* and will be the core of those that will be the cutting edge in the future.

Wars are the mechanism of banks. The money ebbs and flows with each new conflict. Millions suffer and the banks make money off the losers as well as the winners. In the great scheme of things, once the warrior class grew beyond protecting the clan, the accountants and money changers became the new overlords. It's pathetic that our moral fiber has degraded to where we worship wealth and fame more than we guide our lives with honor and integrity. My companions and fellow travelers pride themselves on their word and go to fatal lengths to maintain it and their honor.

We begin this new millennium with our country in jeopardy, the corruption emerging from the quicksand called Washington overwhelming the senses. We are under house arrest in our own country, with rumor and innuendo stalking the land like direwolves. "The World Turned Upside Down" was the tune that Cornwallis ordered played as he marched out and surrendered to the Virginia planter and our Republic was born. I hope we don't play the same tune. We can't help but be fervent about this country, we have bled for it, watched our friends die for it, and in our apathy, let tyranny take hold.

* * *

If you are attracted to the flame, try and carry a parachute. That is the nature of what we do. By choice we took this path. We have lived our lives, sometimes not so wisely but we have lived them. Our contemporaries are the same. We have a saying in Special Forces—it isn't a job description, it is a way of life. The new generation shows promise. If that is true, then there is the chance for a legacy. Watch out for old men in a young man's profession. They survived!